TIP OF THE
SPEAR

OUR SPECIES AND TECHNOLOGY AT A CROSSROADS

TIP OF THE SPEAR

JIM A. GIBSON

Copyright © 2018 by Jim A. Gibson.

All rights reserved. No part of this publication may be reproduced, distributed or transmitted in any form or by any means, including photocopying, recording, or other electronic or mechanical methods, without the prior written permission of the publisher, except in the case of brief quotations embodied in critical reviews and certain other noncommercial uses permitted by copyright law. For permission requests, write to the publisher, addressed "Attention: Permissions Coordinator," at the address below.

Jim A. Gibson/JAGYYC
8240 10th ST SW
Calgary, AB T2V1M9
www.JimAGibson.com

Interior design: Typeflow

Ordering Information:
Quantity sales. Special discounts are available on quantity purchases by corporations, associations, and others. For details, contact info@JimAGibson.com.

Tip of the Spear/Jim A. Gibson.—2nd ed.
ISBN 978-1-7751289-2-2

For my mother, who gave me my life-long love of books, Drambuie and a good debate;

For my father, who never forgot that it's how you treat everyone that matters.

Men go forth to marvel at the heights of mountains and the huge waves of the sea, the broad flow of the rivers, the vastness of the ocean, the orbits of the stars, and yet they neglect to marvel at themselves.

Saint Augustine of Hippo

contents

Introduction: *The Library* — 1

PART 1: MEET THE TIP OF THE SPEAR — 15
  1. Introducing the Three Laws — 17
  2. **Law #1:** *The Slope of the Disruption Curve is Dramatically Increasing* — 19
  3. **Law #2:** *The Genie and the Bottle* — 35
  4. **Law #3:** *Our Linear Systems of Organization* — 51

PART 2: MEET THE TAIL — 61
  5. **Inequality:** *The Tip Moves Far Ahead of the Tail* — 63
  6. **Inequality:** *Challenges in the Developing World* — 73
  7. And one more thing… — 83

PART 3: THE JOURNEY OF TWO GIRLS — 95
  8. **Olivia:** *The Great Social Experiment* — 97
  9. **Olivia:** *An Education System Disconnect* — 113
  10. **Olivia:** *Disruption and the New Way of Work* — 117
  11. **The Lessons of Olivia:** *Actions Now* — 123
  12. **Mervis:** *Education Drives Everything* — 127
  13. **Mervis:** *The New Technology Revolution* — 133
  14. **Mervis:** *Building the New Africa* — 139
  15. **The Lessons of Mervis:** *The New African Experiment* — 145

PART 4: TOWARDS A NEW SOCIAL CONTRACT — 149
  16. A New Definition of Innovation — 151
  17. The Rainforest and the Human Collective — 159
  18. Social Contracts for the Rest of Us — 171
  19. Skills of the New Business Leader — 207
  10. Lead, Follow, or … — 221

Endnotes — 227

ACKNOWLEDGEMENT

When you write your first book after 57 years on the planet there are many people to acknowledge and thank. The single biggest factor in writing a book is—no surprise—the love of reading. My parents gave me this, starting with a houseful of books and a vintage version of the World Book encyclopedia that was essential in slaking my thirst for knowledge as a kid. It all followed from there and I thank them every day for the gift of curiosity.

Geoff Moore, my friend of over 40 years and author of three fiction titles took on the roles of grumpy wordsmith, finder of my voice and holder of the red pen. In Geoff's case, literally a red pen. He brought back memories of my Grade 11 Jesuit English teacher scrawling, "Vague. Re-write" with a red circle across an entire page in my early drafts. I thank him for the lesson of humility in the craft.

To my readers, editors and all of those who have given their time, energy and love, my deepest thanks for giving me the courage and encouragement to continue and for their gentle corrections that magnificently walked that tough line of peers and family critiquing someone close. I thank them for helping me make this book more than simply a completed task and into one that will stand the test of time and of which I will be everlastingly proud.

My wife Cynthia knew instinctively that her "Dutchness" was best used in keeping me on task rather than being another editor. This will be her first read of the book. I am beyond grateful for the space, time and love to be allowed to take this most selfish of journeys.

Finally, to Olivia and Mervis—who you are about to meet—I thank you and your cohorts. I think about and worry about you every day. I will make it the focus of my "third period on the ice" to ensure that we begin the conversations that will allow you and your generation to slay the dragons of inequality and create a planet that we know our species is capable of building.

Jim Gibson
December 2017

TIP OF THE
SPEAR

INTRODUCTION
THE LIBRARY

This story begins in a public library[1] in my hometown of Calgary, Alberta, Canada. Though it is a lovely spot, I am not a frequent guest. The last library I regularly visited was at the University of Toronto—St. Michael's College, to be precise—many (many!) years ago. I would do what most students did at libraries: spend an hour or so studying and the rest sleeping. Sometimes to shake off the night before but mostly from brushing up on the mind-numbing microeconomics theories that for the most part had me wondering, "Who made this stuff up?" Nod off.

The experience of this mid-summer day in 2016, however, is very different. I have come here to find a quiet place to focus, think, research and—hopefully—write. But something else is happening. I am getting an unintended up-close view of the cross-section of humanity that visits the 2016 version of your humble community library. And it's connected to the story I am about to tell.

This book is called *Tip of the Spear*. Its central thesis is that we are quickly coming to a time in human evolution that requires some thoughtful and connected human responses to big problems. The tip of this spear is the white-hot technology that will change the very nature of our planet in the years to come. This tip is moving very quickly away from the butt of this spear—the inequality and the unleveled playing field of our world and the human lives being affected.

It is, ultimately, a story of hope and possibility. But it is also a shot across the bow, a recognition of our indifference and inattentiveness. It is a cautionary tale. It stems from my belief that we are about to be confronted by changes that will rock our collective and individual understanding of the world. I have been on the cutting edge of technology for 30 years in the various start-ups I have built and sold, in large enterprises where I have been an inconsistent and reluctant employee and in my teaching and engagement with the technology ecosystems here in Canada, the US and the UK. I have been a witness to and active participant in the shaping of the technology that has brought us to where we are today.

Much more on that in the coming pages, but now, back to the people in this library.

There is Late Middle Age Man contemplating life, looking out the window. There is Out of Work Oil Worker struggling at the shared computer. There is Distracted University Woman writing a paper but texting and looking at Facebook far too often. There is Retired Physically Challenged Lady who looks like this is the place she comes to feel alive. Homeless Man looks nervously at the newspaper rack. Security Lady patrols the shelves.

Of course this is a poor cross-section for demographic analysis or frankly anything remotely scientific. But what it does is force me to confront the fundamental question all writers face (or should face) when they begin with a blank page, namely, "Who am I writing this book for?" More in the moment, I ask, "Given that I know what the tip of the proverbial spear looks like, what will happen to these people in the next five or ten years?"

The subject matter is complex. The infamous 1%—who are now popularly vilified as the elite—need to understand that the technologies they are shaping through investment, work, research or exploitation carry with them extraordinary baggage and unintended consequences. The 99% need to appreciate what is coming and—somehow—manage themselves accordingly.

I realize as I look about the quiet room that if I try to write for both, I will serve neither very well.

But the problem hits me: Even if I settle on the "voice" and "audience," I am told by many that people—of all stripes—don't read anymore. It's all high bandwidth now, right? Blast the visual cortex with as much input as possible. Multi-task the world around you.

Maybe I should call this, "The Last Book You Will Ever Read…". Or perhaps I should just record a couple of Snapchat stories and YouTube snippets (less than two minutes, of course) to get all of this across.

I stare warily at my full and complete table of contents. There is so much to be said. It is a story of thought and of some length. It can never be captured in simple vignettes.

That's when it hits me. Who isn't here?

Thirteen-Year-Old Teenager (and their tribe of plus or minus three years).

As I sat contemplating the future lives of people in the library who are—for good, bad or indifferent—the sum of their decisions and will carry out the rest of their lives on relatively the same vector, I realized that the decisions my cohorts make in the very near term will decide the shape of the world for the pre-teens and young teens of this world.

I realized I needed to understand their viewpoint because it is in their immediate lifespan of the next 10 to 15 years that all of this needs to be worked out. A lot normally happens from ages 13 to 25. But the decisions that the world needs to make in this timeframe will, I believe, change the course of our species, our planet and our reason for being here at all.

My mind flashes back to a game-changing event in my life and—more importantly—in the world around me: Ten years ago the very first iPhone from Apple was released to an unsuspecting world.

Ten years ago. A blink of an eye to the adult population. A lifetime—literally—in the life of a 13-year-old. In the intervening ten years the world has become social and mobile. In an instant, we changed the way we work, live and play. We are literally altering the way we interact with the physical and digital world.

This book is not about technology predictions yet this timeframe was not lost on me. Twelve years before the release of the iPhone, the world's first graphical browser—Mosaic Netscape—was released. People couldn't have imagined the release of such a product (okay, maybe Alvin Toffler did, but certainly not anyone close to the mainstream of society).

Even in 1995 when Netscape set the world on fire, no one could have imagined the changes that would be unleashed on society because of the broad availability of this software application and its democratization of the world's information laid upon the underlying framework of the Internet developed 25 years earlier.

Therefore, I thought, how can people possibly understand what's coming next?

No one can.

However, there are a few things we can say with almost certainty: During the next 12 years as our 13-year-olds begin their young adult lives, there will be a continuation of the exponential technology march that has been ongoing for 50 years. In 12 years there will be six more "doublings" of technology according to Moore's Law. In these 12 years of exponential growth, we will begin to combine eight or more different technologies that have never been connected, let alone existed before. Young adults will become the first 25-year-olds that have never not had a search engine and will have been carrying a supercomputer in their pocket for 15 years. Teenagers will likely be the last generation needing to have a driver's license. They will live to be 120. Guaranteed. They will probably have to re-invent themselves a dozen times.

But during these intervening 12 years, some very complex decisions will have to be made. The world's resources and profits from the technology revolution have been unevenly distributed. Governments will have to understand and deal with our ability to easily and cheaply manufacture everything from airplanes to body parts. Our genome will be open sourced and our capabilities to fix and alter our body will be staggering.

The very power of the economic engines of the exponential digital economy will be an intimate part of everyday life. Today's youngsters will have the keys to the digital kingdom; will we ask them to help us create this future or send them to their rooms?

In thinking about and preparing to write this book it was clear that its audience was anyone who directly or indirectly influenced, cared about, or who in fact was part of, this coming generation. Our species and technology are at a crossroads. It is mankind's defining time, it will play out in the next 15 years, and there is a way forward.

WELCOME TO THE TIP OF THE SPEAR.

Meet Olivia and Mervis

Before we dig in, meet Olivia and Mervis—two remarkably normal girls; two 13-year-olds about to start the most perilous journey in the history of mankind.

Olivia Browne is a typical Canadian teenage girl. She has just finished grade seven at her local junior high school.

Boys (crush!), girlfriends, *Stranger Things*, the latest fashion trends, soccer and her studies make up her life. In 12 years she will be 25—the age when most of us are well into our first jobs, away from home and absorbing the world around us.

She is constantly told to focus on her studies and get good grades so that she can get into the best schools and get a good job.

And one day she looks at her mother and says, "If I have all of the world's knowledge and all of the smartest people in the world one click away, why should I care how or what I am being taught?"

Olivia has a point. You know in your gut the skills she has been learning for half of her pre-secondary schooling are supposed to lay the foundation for what she will learn next—but will not. The curriculum and learning approaches that await her for the next five years and more would look familiar to her great-grandmother. She will be able to access all of the world's knowledge at her fingertips but won't be able to read because she has long stopped using the slow-paced but fabulously creative muscles of the reading brain. Will it matter?

All of her colleagues have a laptop computer, wicked fast broadband Internet, a supercomputer masquerading as a smartphone. She lives in a country with universal healthcare, a seemingly infinite amount of accessible clean water and an abundance of traditional and unconventional energy sources.

The future while uncertain, seems bright, full of change and possibility.

But there is another reality. Left unchecked her "developed" world is one that social and technological disruption could very well leave behind. With Brexit in Britain and Trump's election in the USA, the capitalist, democratic underpinnings of the developed world are under siege. The very technology that the freedom of capital and expression built has created a new challenge: we increasingly see the world through an always-on filter that grossly limits our world view. It is a world with an economic base that could disappear before her very teenage eyes.

In a very real way Olivia is being prepared by parents who are unknowing of the coming disruption. Those well-meaning folks consume the best of what the innovation economy provide but are largely ignorant of the scope and breadth of the technological and sociological changes coming. The education system to which they entrust their child is—based on current curriculum and approaches—certainly preparing her for a world that increasingly doesn't need her skills. Her parents' cohort—the politicians and leaders—will be too slow to react.

The advent of the smartphone and its cousin the tablet was followed quickly by hand-wringing about the deleterious effects of "screen time." But the impact of these devices has not been fully appreciated, and goes far beyond the usual concerns about curtailed attention spans. The arrival of the smartphone has radically changed every aspect of teenagers' lives, from the nature of their social interactions to their mental health. These changes have affected young people in every corner of the nation and in every type of household. The trends appear among teens poor and rich; of every ethnic background; in cities, suburbs, and small towns. Where there are cell towers, some unironically in the guise of church spires, there are teens living their lives on their smartphone.

All Olivia wants to do now is Snapchat her friends.

Meanwhile, in Malawi

From Wikipedia: "Malawi is among the world's least-developed countries. The economy is heavily based in agriculture, with a largely rural population. The Malawian government depends heavily on outside aid to meet development needs, although this need (and the aid offered) has decreased since 2000. The Malawian government faces challenges in building and expanding the economy, improving education, healthcare, environmental protection, and becoming financially independent."[2]

Malawi has a low life expectancy and a disturbingly high rate of infant mortality. There is a high prevalence of HIV/AIDS, which is a drain on the labour force and government expenditures. A Malawian who survives infancy can look forward to another 50 years of existence.

Traditionally, educating girls was not viewed by families as a priority compared to boys, especially in rural families where the extra pair of hands that a daughter provides at home can help a family survive. Girls shoulder the burden of household chores, including fetching water and firewood, often walking for hours a day. They look after younger siblings when parents need to tend fields or go to market to sell their produce.

Even when girls do attend school, the dropout rates are higher than that of boys. School facilities are often inadequate to provide a quality education; overcrowded classrooms, outdoor lessons interrupted by rain, and perennial teacher shortages mean that girls often don't get the attention they deserve in class, and are not motivated to continue. The dropout rate gets even higher with the onset of puberty.

Meet Mervis Chatha, also 13. She has been sponsored by a unique program called "GirlUp."[3] She sums up the importance of educating girls espoused by the program. "It is important for girls to go to school because when you teach a girl child, you teach the whole world," she says. When women are educated, they are more likely than men to pass knowledge on to their children and insist on schooling. In addition, educated women can earn incomes independently from their husbands, and are more likely to spend money on essential medicines, food and other basic needs for their family.

"I want to be an architect!" says Mervis. "I want to help

build new places and cities in Africa. I pray to God that my family and community will be safe and have all the things that the rest of the country and Africa has."

The future of Mervis and countries like Malawi are both daunting and amazing.

There is a realistic possibility that the scope and speed of change in her world will exceed that of her cousin-in-age across the world, Olivia. Some statistics:[4]

- 700 million people will be moving into African cities in the next 35 years—that means an entire New York City has to be built every six months until 2050.

- Nine out of the 20 fastest growing economies in the world are in Africa.

- Africa is home to one billion people and 200 millions of these are aged 15–24.

- Africa has the fastest growing middle class in the world.

- Smartphone usage is at a tipping point, expected to reach 400 million users in 2020.

Mervis' government is beginning to make fundamental decisions about her education, the role of women in the society and even more fundamentally, the role of the state and its fight against corruption. As the global Internet becomes widely available through initiatives like Mark Zuckerberg's *Internet.org* and others, Malawi and Mervis have a chance to fast-track the cycle of change.

More promising, there will be no legacy of the 60–plus years of experimentation, pollution and failed urban design centered on the internal combustion engine that ultimately saw the emergence of hollowed-out downtowns and smog filled urban centers in the developed world.

Hers will be the first generation of Malawian girls who will not only be educated but will have access to all the world's knowledge. The cultural and societal changes will be monumental and will rival all of what Olivia will have to cope with and more.

But it is also a perilous time for Mervis. Long a victim of colonialism, exploitation and corruption, Malawi has much to do. Decades of progress can be erased in the face of myopic and self-serving governments. Exploitation by new colonial masters—the new technology giants, for example—will deliver the economic and social benefits to a precious few.

So as we take this journey, let's keep in mind both of these girls' lives and seek to understand the implications of the decisions being made at both the tip and the tail of the spear through their eyes.

This story will lead us to two very different visions of the future: how the lives of these two girls, from different parts of the world, will be shaped by the very decisions we will make in the coming years about how we consume, legislate and navigate the coming tsunami of technological change.

A Brief Look Ahead

> "The future has arrived—it's just
> not evenly distributed yet."
> *William Gibson, 1992[5]*

Thinking about my newfound muses in the context of this wonderful and oft-cited quote from Mr. Gibson (no relation, unfortunately), I recalled an important moment in my earlier thinking and research.

I was compelled to ask two simple questions: What do Donald Trump and Ray Kurzweil have in common; and who is Ray Kurzweil?

Donald Trump? Well, we have had to create a separate Internet just to handle the digital spew written about him. I'll get to him in a moment.

From his biography: Ray Kurzweil has been described as "the restless genius" by the *Wall Street Journal*[6], and "the ultimate thinking machine" by *Forbes* magazine[7], which ranked him #8 among entrepreneurs in the United States, calling him the "rightful heir to Thomas Edison." PBS selected Ray as one of 16 "revolutionaries who made America," along with other inventors of the past two centuries[8]. He is considered one of the world's leading inventors, thinkers, and futurists, with a 30-year track record of accurate predictions.

Two of Ray's books have defined an age: *The Age of Spiritual Machines* has been translated into nine languages and was the #1 best-selling book on Amazon in science. *The Singularity Is Near: When Humans Transcend Biology* was a *New York Times* bestseller, and has been the #1 book on Amazon in both science and philosophy. It predicts a future point in time, "The Singularity," when machine intelligence surpasses human intelligence. An amazing and frightening read.[9]

How can two humans as impossibly different as Trump and Kurzweil have anything remotely in common?

It was a strange occurrence in my daily reading when I came across an *Economist* magazine[10] treatise on the state of Artificial Intelligence followed immediately by their artful bashing of Mr. Trump that left me wondering, "Is there a connection here?"

As I thought about it, I finally put into focus something

that has been concerning me for a while: the growing distance between the promise of the future and those driving it forward, and what I call the future disenfranchised, those with very little chance of benefiting from this future. They are found in the rural hinterland of America where manufacturing has long left, and cruel fiscal choices have laid waste to entire generations. They are found in the staggering urban slums of Lagos, Nigeria where the potential for hydrocarbon-fueled riches for the entire population has fallen victim to systemic corruption.

They are also found in our hearts. We connect across the world to the best of what makes us human but ignore the inequality and suffering in our own backyards. *The Future Disenfranchised* might very well have been the title of this book.

But as I contemplated this deeply troubling future, I came to understand that there was a pattern. And with this pattern, I began to see my way through it.

Three Simple Things

I sketched out a whole bunch of theories and frameworks that tried to explain this. I met with, drank beer with, and debated with some very smart colleagues from around the world. In the end it came down to some very simple things that I know to be true—Gibson's Three Laws of Disruption:

1. The slope of the technology disruption curve is dramatically increasing.
2. The technology "genie" never goes back in the bottle.
3. Our linear systems of human organization are unprepared for sustained exponential change.

And there it is. Our future will be decided—soon—by how well we cope with and address these forces that are, as you will soon see, fundamentally at odds with each other. Individually, these three observations that I have elevated to "laws" are at first blush mostly self-evident. Taken together, however, they present a set of circumstances that connects the human animal across all of time. The difference today is that the velocity of change is met head on with the impossible struggle to absorb change across complex and intractable societies and institutions.

Our use of the technologies we have precariously balanced are now poised on a blade's edge. Either the discussions and actions started here (and elsewhere) begin to level the playing field in areas such as governance, finance, and education, or we use these extraordinary technologies to create further concentrations of wealth and increased inequality. Either we use our newly connected planet to have real conversations or we only hear the rabble at the gate.

The "rabble" is increasingly making our citizens, companies and governments nervous. We see it in the Brexit vote, the rise of the American disenfranchised via Donald Trump,[11] and the persistent slant of the playing field across the world of finance, education and opportunity. I also see it in my city as it struggles with a new economic order and model. I see it in the eyes of the newly unemployed. Finally, I see it right here in the municipal library.

It seems a cruel joke to me that just as we finish wiring up over half the planet to be able to connect the best in us— our ideas, hopes and innovation—the chasms start to show. Slowly at first, but inevitably in an exponential shift.

Our linear minds and linear institutions simply cannot keep up.

The purpose of this book is to use these "truths" to dissect what is going on around us and deliver both an appreciation of the challenge (Parts 1 and 2) and a way forward (Parts 3 and 4) as I dissect these three themes in detail.

In summary, my hope is that *Tip of the Spear* does three things. Firstly, it helps my colleagues, companies, individuals and institutions to really understand the scope and speed of change that is coming. Secondly, it will prepare them for a world that must understand that the technology genies are not going back in the bottle. Finally, it shows that in our slow-moving, intractable and inflexible organizations and institutions (including democracy itself), there is an incredible new opportunity—likely an imperative—to think exponentially and act differently.

My work is to take one individual, one company, one institution, one linear model at a time and help make the change happen. Let's dig deeper and understand the root of these disruptions. We begin with Part 1, humbly called 'Gibson's Three Laws of Disruption" and, as the other Gibson noted earlier in the chapter, we will discover for ourselves that the future indeed is here and certainly not evenly distributed.

PART 1
MEET THE TIP OF THE SPEAR

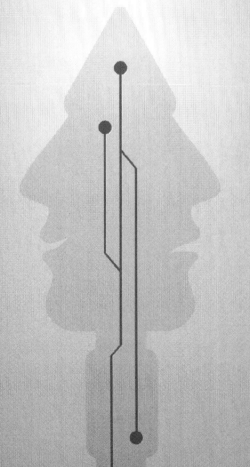

chapter 1

INTRODUCING THE THREE LAWS

As part of my work and research over the past decades, generally, and the focused efforts in the past two years interviewing and reading the leading thinkers in the technology disruption space, I have summarized my understanding of what's coming as Gibson's Three Laws of Disruption:

> **GIBSON'S THREE LAWS OF DISRUPTION**
>
> **1**
> THE SLOPE OF THE DISRUPTION CURVE IS DRAMATICALLY INCREASING.
>
> **2**
> THE TECHNOLOGY GENIE NEVER GOES BACK IN THE BOTTLE.
>
> **3**
> OUR LINEAR SYSTEMS OF HUMAN ORGANIZATION ARE UNPREPARED FOR SUSTAINED EXPONENTIAL CHANGE.

These three observations are the basis of understanding the Tip of the Spear and why it is important. They have been formed through my participation in the world of technology for over 30 years and by being in a front-row seat of understanding how technology change affects the average person. They have also been informed by connecting the dots of a set of macro forces that—you will soon see—have been brewing for many decades.

We will look at each of these laws in detail and then turn the spear around and understand the reality of how fast the tip of this spear is moving ahead of the rest of humanity. We will then provide a look into the future by taking a hypothetical journey with our two 13-year-old girls. These scenarios helped me understand and empathize with the implications that these three laws will have for the very cohort that will be most affected.

This empathy will be important to the final part of our journey, as we look to what we can do today as we come to grips with our species and technologies at an existential crossroads.

To begin, let's dig into the first law and understand what exactly is happening at the white-hot tip of the spear.

chapter 2

LAW #1

THE SLOPE OF THE DISRUPTION CURVE IS DRAMATICALLY INCREASING

What started out as an obvious "first law" for me has proved to be the most controversial and the most misunderstood. I cannot tell you how many people are genuinely confused or underwhelmed by it. One explanation is that we have been lulled to sleep by the idea that since technology is always changing, the current trend of technological disruption feels like all the others. In other words, technological advance, yawn!

For example, in the recent past, the introduction of the masses to the Internet with the release of the first graphical browsers was quickly followed by the rise of mobile computing, first with the introduction of the Blackberry and then with the iPhone. Then we had the rise of social media with Facebook and the like. If you are a history buff, we think of change as the arrival of the Renaissance with the invention of moveable type, the Industrial Revolution with the arrival of steam or the mass production era when Henry Ford released the Model T. We look back on Kitty Hawk in 1909 and project forward to Apollo 11 in 1969.

Change is everywhere, and for most people, technology change always follows an upward sloping line.

But the first law is subtle: It isn't simply that the technological change curve is upwardly sloping to the right. It is that the slope of said curve is getting steeper. And that means that "change" is very, very different this time.

Here are three reasons why:

1: *Today's Change is Exponential*

To be a math nerd for a moment, if the slope of the curve is increasing it means that rate of change is increasing. Even my most advanced technology friends and colleagues have to think twice about what this really means. In layman's terms the hill (up or down) is getting steeper and steeper every step you take. Worse, like the story of Sisyphus, at some point it becomes impossible to move forward as the curve becomes perpendicular.

In my research interviews, most people really do appreciate that things are moving quickly when presented with the overwhelming evidence of the scope, speed and breadth of change. But when I mention that the very slope of this exponential curve is increasing, then it gets a little dicey. Never mind saying that that first derivative of the exponential curve is increasing. (I don't really talk about that!)

So let's go back to basics for a second. Of course we have to talk about Moore's Law.[12]

The last thing we need, you say, is another futurist quoting Moore's Law. Every book or article on technology and innovation uses it. But we have to understand it—perhaps as a repeat for some of you—to drive home the point that it underlies everything we are about to experience. While there are many ways to express it, the basis for Moore's Law is the empirical observations that price-performance of computing—that is, the ability of the computer's microprocessor to perform raw calculations per unit of cost—is doubling every 18 to 24 months. After celebrating 50 years at the heart of technological innovation, the "Law" has become a driver of innovation: innovation is being invested in to keep the trend going. It's a pretty big deal.

But let's remind ourselves of the basic math at work here. The curve you traditionally see on a "Moore's Law" slide is a straight line. The reason: It is plotted on a logarithmic scale. Instead of the vertical axis having ticks of equal distance equaling the same amount (10, 20, 30, etc.), the equal ticks jump by powers of 10.

Typical examples look like this:[13]

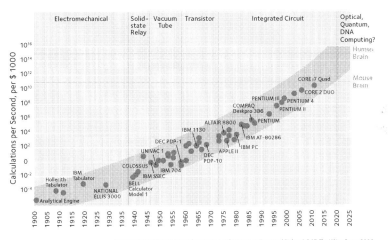

Source: Ray Kurzweil, "The Singularity is Near: When Humans Transcend Biology", P.67, The Viking Press, 2006. Datapoints between 2000 and 2012 represent BCA estimates. © BCA Research 2013

The point is that unless you are a math person, this is NOT the way we normally look at line charts. It is important to remind ourselves that a linear scale looks a lot different—something more like this:[14]

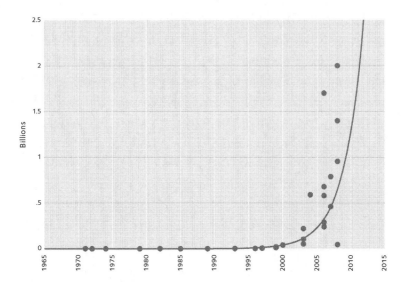

Why show the linear scale? For two reasons: First, we get seduced by the gentle upward sloping curve of the exponential scale. We get hypnotized as we believe that change is an upwardly sloping slide. The second reason is to visualize and understand the "elbow" or "knee" of the curve. When it takes off, it really takes off!

Looking at the graph of Moore's Law shown linearly we see this effect. We are well past the "elbow" of the curve in terms of raw computing power when we look at the capabilities of our raw computing power. In other words, things are really about to explode.

For our fundamentally "linear" brains, this sure grabs our attention, doesn't it? That's why I remind people that we need to understand this change visually in terms we can easily appreciate and understand. There is simply no doubt that the linear expression of this exponential function makes us appreciate that stuff is moving ahead quickly—and (I hope) understand that the RATE of change is increasing.

But there is much more to this.

2: Technology Costs Drive to Zero

Where things really get interesting—especially to those investing in the companies creating new technologies and organizations struggling with business models that are being completely eroded—is when we flip these curves and look at what this exponential capability does to the cost of a "unit" of computing.

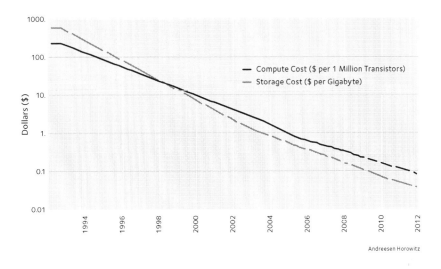

What emerges is the mind bending question, "What happens when the cost of computing (and associated technologies) approaches zero?"[15]

That is the end result of this, isn't it? As we approach the limits of increased price/performance, computing and technology is becoming essentially free. Note, this is about *cost* of technology (the sum of all inputs to create something) not the *price* of technology (what someone pays for something).

This is what Peter Diamandis eloquently talks about in his seminal book, *Abundance: The Future is Better than You Think*.[16] Diamandis takes this basic set of observations to their logical conclusions. If we believe the trend will continue, then many of the input costs for technology will eventually drive to zero. By definition this makes them abundant and our models of planning and living in scarcity fall away. Since the Industrial Revolution beginning in the early 1800s, the inputs of production—land, labour, capital and entrepreneurship—and the laws of supply and demand underlying classical economic theory have been governed by the simple principle of scarcity. Maximizing the inputs and pricing goods and services accordingly has driven our capitalistic society since that time.

In a world where inputs are following exponentially decreasing *cost* curves at the same time as exponentially increasing *capability* curves, our understanding of the basic foundations of economics is dramatically altered. This simply isn't something for which our slow-to-evolve cultural brains are wired!

Assuming we can get our heads around the idea of a future built on abundance, there is something else that has gotten my attention: the combination of exponential technologies.

3: Combinations Have Consequences

I have long said that Moore's Law—while absolutely fascinating and a fundamental driver of technological innovation—is just one of the curves to which we need to pay attention.

Digital communication speeds, digital storage, genome sequencing and a host of other technologies are actually following similar arcs. In the case of some of these, the slope of the curve of change is even steeper!

Individually, this is exciting and amazing. But it is actually these combinations of exponential technologies that Diamandis, Kurzweil and others at the Singularity Hub talk about that make it particularly amazing (or frightening)—but definitely important.

In a set of seminal articles written by global experts in their respective fields, Diamandis and Kurzweil identify and discuss eight technologies that are driving the future:[17]

1. Computation (Moore's Law)
2. Internet of Things (Sensors & Networks)
3. Robotics/Drones
4. Artificial Intelligence/Machine Learning
5. 3-D Printing
6. Augmented and Virtual Reality
7. Materials Science
8. Synthetic Biology

I would add a couple more:

9. Blockchain and Digital Identity Systems
10. Brain and Cognitive Sciences

The point here is that, as more of these exponential technologies are combined, you start to appreciate that the changes we are seeing today are completely different from those of previous generations. It is as though we have the Renaissance, Industrial Revolution and the Information Age all happening simultaneously. Furthermore, the natural governors of previous eras—time and space—do not exist anymore. The digital infrastructure that has been laid in the past 30 years, enables the global collaboration and sharing of the science and commercialization instantly and—with some social media driven annoyances—transparently.

And this is where the action is! As we begin to be able to combine sciences and the smart human beings who lead the research and investment, we are creating not only new inventions but new ways of working. As Aaron Frank notes in his profound essay, "The World Depends on Technology No One Understands," the "edges" are where it's at—especially as we connect physical and life sciences together with computation and machine intelligence.

The implications are not only in the potential inventions and new technologies that will emerge but the impact on learning institutions all over the world. Long a place of siloed departments pitted against each other for access to scare resource, facility and attention, our universities and public research fields are—slowly—starting to realize the power of the intersections of fields of study.

Frank notes,

> ...A great deal of substantive science is being done now where the life and physical sciences intersect, and even more transformative research is on the horizon. While the potential benefits for society are profound, realizing that full potential will require significant changes in how we educate, train, and support those undertaking this research.
>
> Intentionally or not, our system pits one scientific area against another in competition for a limited pool of resources. To obtain the benefits from research at the intersection of the physical and life sciences, it will be necessary to overcome these obstacles and to create a scientific structure that truly reflects the scientific needs and opportunities of twenty-first-century science.
>
> Breakthroughs occur when scientists from a variety of disciplines either individually or collaboratively work on important interdisciplinary and multidisciplinary problems. Therefore, we need a new generation of scientists with both rigorous disciplinary training and the ability to communicate and work easily across disciplines.[18]

In summary, the additive factors of combining exponential technologies and the disappearance of natural governors of time and space creates what I call **Gibson's Law of Innovation:**

> ## GIBSON'S LAW OF INNOVATION
>
> COMBINATIONS OF EXPONENTIALLY CHANGING TECHNOLOGIES WILL RESULT IN AT LEAST ONE UNINTENDED AND UNFORESEEN NEW GLOBAL TECHNOLOGY AFFECTING A BILLION PEOPLE OR MORE EVERY FEW YEARS.

As we begin to create ever increasingly capable machines that in turn are able to create exponentially new machines (for example, a cloud-connected, collaborative gene sequencing machine talking to all other cloud-connected gene sequencing machines using machine learning-based communications, storage, fabrication and computation powers with rapidly diminishing marginal costs), we will have a new order of magnitude of change. Most if not all of it will be unintended and without precedence.

In other words, the innovation hill we have begun to climb is really getting steep. The first derivative is very positive. I particularly love this quote from Peter Diamandis' book *Bold: How to Go Big, Achieve Success, and Impact the World*, where he imagines looking back on our world from a future time:

> By 2020, a chip with today's processing power will cost about a penny," CUNY theoretical physicist Michio Kaku explained in a recent article. . ."which is the cost of scrap paper. . . . Children are going to look back and

wonder how we could have possibly lived in such a meager world, much as when we think about how our own parents lacked the luxuries—cell phone, Internet—that we all seem to take for granted.[19]

Digital disruption now becomes that once-in-a-species opportunity to create things we never even dreamed about, increasingly supported by stuff that is becoming free. This is an extraordinary time to be a creator of new things, services and businesses.

Exponential increases in solar array cells and energy storage costs drive energy cost and storage to zero. 3D printing drives car manufacturing and infrastructure to zero. Autonomous cars drive cost of driving (accidents, traffic, waiting, purchases, insurance, etc.) to zero.

In Olivia's lifetime, no license, no car ownership, no insurance costs, no gas costs. No friends dying at the hands of drunk drivers.

Mervis will see a transition from rural to urban life that completely skips three or four generations of roads, crumbling infrastructure and a communications framework that challenges systemic inequality and potentially levels the playing field.

To my business colleagues in almost every imaginable business arena: change will come fast. I believe it. I have lived it. I make my living understanding the implications, causes, and strategies for these technologies. And I have never experienced such clarity assessing the impact of these relentless waves of change. The tide cannot be turned back.

So do me a favour: Read the latest posts[20] from the Singularity Hub blog about the top eight technologies, their current state and the impacts in the next five years. I can assure you that after reading you will appreciate that we are on the "elbow" of the exponential curve—that the technological change is just getting started. Check out the World Economic Forum's Top 10 Emerging Technologies of 2017.[21] and Chris Dixon's "What's Next in Computing."[22] Or if you want a more humorous but equally accurate and engaging post, read Tim Urban's funny, thoughtful and scary reality check on Artificial Intelligence.[23] If you want to know what's coming in the world of financing, trust and identity, read Don Tapscott's book on the Blockchain.[24]

Some excerpts to whet your whistle (or in case you don't feel like reading more): As the Singularity Hub notes,

> *an expert might be reasonably good at predicting the growth of a single exponential technology (e.g., 3D printing), but try to predict the future when AI, robotics, VR, drones, and computation are all doubling, morphing and recombining. You have a very exciting (read: unpredictable) future....*[25]

To quote Kurzweil: The Law of Accelerating Returns:

> *The first technological steps—sharp edges, fire, the wheel—took tens of thousands of years. For people living in this era, there was little noticeable technological change in even a thousand years.*
>
> *By 1000 A.D., progress was much faster and a paradigm shift required only a century or two. In the 19th century, we saw more technological change than in the nine centuries preceding it. Then in the first 20 years of the 20th century, we saw more advancement than in all of the 19th century. Now, paradigm shifts occur in only a few years' time. The World Wide Web did not exist in anything like its present form just a decade ago, and didn't exist at all two decades before that. As these exponential developments continue, we will begin to unlock*

unfathomably productive capabilities and begin to understand how to solve the world's most challenging problems. There has never been a more exciting time to be alive.[26]

After you have read enough, you cannot help feeling exhilaration but also a deeper sense of unease that there are going to be a lot of folks who will be profoundly affected by these changes. Many don't yet have a clue about what is coming and only a very select few, the tip of the spear, have had the luxury of time, intelligence, education and capital to anticipate, plan for, and obtain the keys to this new digital kingdom.

But here is the real challenge for us as we absorb these opportunities and challenges: our species' tendency to overestimate technology in the short term and underestimate it in the long run.

Estimating the Future

Without question, the slope of the curve of disruption is indeed increasing—and fast. However, very few of us are aware or understand the scope and speed of these changes. There are many theories as to why, but I think there are two things that we do as a species—and have consistently done over the millennia—that cause this blind spot.

First, we consistently overestimate the impact of new technology in the short term. We miscalculate our capacity, desire and ability to adapt to the changes brought on by technological shifts. We forget that organizational and cultural inertia are large and powerful flywheels. Further, when guarantees of a better future do not happen at the speed promised by the prognosticators, we develop technology amnesia. We then consistently underestimate the impacts of technological change in the long run. Only a very select few people or companies actually do connect the dots—and in the technology world, to the observant, go the spoils. We have created an innovation and capital system called the venture capital business that with increasingly few expectations rewards a single winner (and of course the cabal that funds them).

For a classic example of this, look no further than Artificial Intelligence. In 1957 the economist Herbert Simon[27] predicted that computers would beat humans at chess within 10 years.[28] (It took 40.) In 1967 the cognitive scientist Marvin Minsky[29] said, "Within a generation the problem of creating 'artificial intelligence' will be substantially solved."[30] Simon and Minsky were both intellectual giants, but they erred badly. Thus it's understandable that dramatic claims about future breakthroughs met with a certain amount of skepticism.[31] Fast forward to the summer of 2017 when the smartest minds in the world were debating to what degree the impact of AI will be an "existential" crisis for mankind. In other words, we now underestimate the power and scope of the AI and Machine Learning revolution.

The concentration of wealth, talent and capital that happens as a result increases, and the potential for benefit for the entire planet falls to the notorious uneven capitalistic view of profit distribution. Open Source solutions be damned![32]

Second, we are showing a disconcerting capacity to trivialize the use of the tools of the gods—that is, a global, interconnected world of knowledge and understanding that would be absolutely mind boggling to someone born 100 years ago. We consistently use our tools to advance the worst of what makes us human. While there are many who remind us that pornography and the adult film ecosystems are often the first users of the advanced technology, this is the least of my worries.

We use a connected world to narrow our world view rather than broaden it. We increasingly use the social tools and the comment sections of online forums to anonymously troll our fellow humans. We are creating a shrillness online that is forcing many of the best and brightest to flee.

Much, much worse, our technological concentrations (see point above) are forcing the best of the advancements through the rinse cycle of self-interest. We have seen hints of this with the advancement of critical pharmaceuticals, for example, held back from broad distribution due to cost and artificial scarcity. Imagine when the genome mapping and synthetic biology advancements begin? What happens then? The specter of "fake news" and the "weaponization" of its soldiers—the social media "bots"—threatens democracy itself. "Truth, it has been said, is the first casualty of war."[33]

What war is this, I shout?

And there is one more thing: Technology advancement, either measured as increasing capabilities or decreasing costs, is not the only exponential curve. Perhaps some of the most eye-popping change is with the physical state of our planet and our impact on it.

Here is one of my favourite summaries from the blog Climate Etc., that connects man-made CO_2 emissions to temperature anomalies:[34]

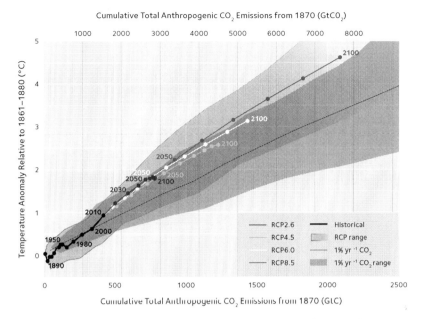

From Nick Lewis in the blog Climate Etc. explaining the direct relationship between Cumulative total emissions of CO2 and global mean surface temperature [GMST]

And of course everyone's favourite: Human population growth for the past 2,000 years:[35]

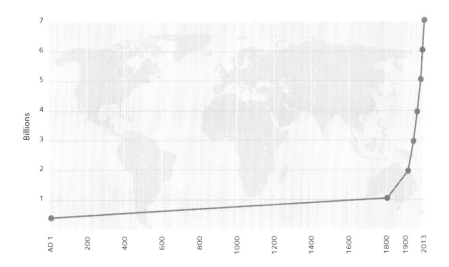

I will be exploring many of the challenges that all of this

data represents throughout the remainder of this story and I will focus much of the discussion on the positive changes happening in pockets around the world, so don't despair. But this section is a cautionary tale.

As Einstein said, "Knowledge of what **is** does not open the door directly to what **should be**."

And that leads us to the second Law of Disruption.

chapter 3

LAW #2
THE GENIE AND THE BOTTLE

Gibson's Law #1: The slope of the technology curve is increasing and that means change is happening at a pace never seen before. Check. Got it.

Gibson's Law # 2 is where it gets serious.

Simply stated: Never in the history of the intelligent human species have we ever invented or discovered something new and then decided *as a species* it was not moral, useful or valuable to someone and subsequently discarded it.

Never. It is simply not in our fundamental nature. And I won't argue that it should be.

This story is ancient and archetypal: Eve in the garden of Eden eating the apple of knowledge; Prometheus creator of man, stealing the fire to protect humans.

Away from the myths, our recorded history begins with the discovery and taming of fire.

From Stephen J. Pyne's essay, "The Fire Age:

The species that won biggest was ourselves. Fire changed us, even to our genome. We got small guts and big heads because we could cook food. We went to the top of the food chain because we could cook landscapes. And we have become a geologic force because our fire technology has so evolved that we have begun to cook the planet. Our pact with fire made us what we are.

We hold fire as a species monopoly. We will not share it willingly with any other species. Other creatures knock over trees, dig holes in the ground, hunt—we do fire. It's our ecological signature. Our capture of fire is our first experiment with domestication, and it might may well be our first Faustian bargain.[36]

I can imagine that when fire was first discovered it came with a terrible cost. Some of the elders of the gathered clan who had lived before the ability to control heat and light saw it only as means to burn and destroy.

Flint and stone tools followed next. Sharp edges. Massive improvement to productivity for procuring and preparing food. Weapons of potential lethal violence.

The wheel. Mobility and productivity; instruments of war.

The printing press. Spread of human knowledge; the control of the collective thought and the ever-present "fake news" problem.

Alfred Nobel invents dynamite and new weapons of war; the peace prize bearing his name is created along with the ultimate irony of the modern era.

Of course you say that there have been staggeringly blind and terrifying moments in history where human progress and human innovation were buried deep under the guise and weight of religious fervor, political expediency or commercial monopoly practice. Or alternatively, the morals of the day decided for one reason or another that certain advancements passed beyond our red line of what is "right" and we chose, using all available political and social tools, to ban them, politically or ethically.

Let's look at one such: Copernicus and the heliocentric model of the universe.

Nicolaus Copernicus—who many of us studied in school—first postulated the then profound heresy that the Earth was *not* the center of the universe. According to his observations and study, the Earth travelled around the true center, which he believed to be our Sun. The mathematician and astronomer stunned the Renaissance world by creating such a model (though one similar had been put forward eighteen centuries prior by Aristarchus of Samos)[37].

The publication of Copernicus' model in *De revolutionibus orbium coelestium* (*On the Revolutions of the Celestial Spheres*), just before his death in 1543, was a major event in the history of science, triggering the Copernican Revolution and making an important contribution to the Scientific Revolution.

The response to his work from all sectors to his work was initially ignorance and a collective yawn. But, more importantly, it became highly controversial and got a lot of people in trouble. One of those people was the Renaissance astronomer and polymath, Galileo Galilei. The Church was not amused. On the orders of Pope Paul V, Cardinal Robert Bellarmine gave Galileo prior notice that a decree was about to be issued, and warned him that he could not "hold or defend" the Copernican doctrine. The corrections to *De revolutionibus*, which omitted or altered nine sentences, were issued four years later, in 1620.

In 1633 Galileo Galilei was convicted of grave suspicion of heresy for following the position of Copernicus, which according to the powers in the Church, was fundamentally counter to the true sense and authority of the bible and scripture and was placed under house arrest for the rest of his life. The Church was trying to put the genie back.

But the book—which had been printed because of the

invention of the equally revolutionary printing press back in the 1400s—was out. The theory and the book took 300 years of abuse and Catholic Church ridicule before 1835 when it was finally removed from the restricted list. But by then the technology of telescopes and the precision found in the math of Copernicus and those that followed had made their way into the world. The technology and science genie almost destroyed the Church and heralded in a new age of science and discovery. The genie though battered and bruised could never be forced back.

In our more recent past, a new controversy emerged. It has been a staggering 75 years since the Manhattan Project and the creation of the first atomic bomb. On the ICAN (International Campaign to Abolish Nuclear Weapons) website, there is an extraordinary table that lists the entire history of our interaction with nuclear weapons—from the establishment of the Manhattan Project in the early 1940's to the explosion of the first two bombs, all the way through to today as we continue to struggle with the treaties around the word and rogue states.

I reprint the entire timeline[38] in the notes for a simple effect: When we review it in its totality we recognize that a single technology—arguably mankind's first existential crisis—remains a fragile (at best) at détente with our species. One technology.

More importantly, try as we might, we CANNOT put it back in the bottle. It haunts conspiracy theorists and doomsday prophets, it has affected geo-politics for decades, and continues to wreak havoc via North Korea, failed states and potential acquisition of nuclear raw material by terrorist groups. Try as we may, the genie won't go back in the bottle.

Now that we've explored some of the genies of the past, let's jump forward to two new existential crises facing the human race: the debate over the potential for harm of

Artificial Intelligence and most recently, Human Genome Editing.

AI: *Mankind's Last Invention?*

From my perspective, Artificial Intelligence (AI) has become the "canary" in the coal mine for the fate of the human species as it meets the technology crossroads. AI is also a classic example—as discussed earlier—about how we overestimate technology in the near term and underestimate its impact in the long run. AI had its first heyday in the late 1970s and 19080s, only to fall in to what was called the "AI winter." But it has most certainly returned in force over the past five or so years.

Ray Kurzweil's singularity, which we discussed in this book's introduction, is essentially the transition state through which the capability of artificial intelligence systems will pass. As Nick Bostrom, one the leading thinkers and researchers in the field of AI, notes in his paper, *Policy Desiderata in the Development of Superintelligent AI*:

> [T]he development of "Superintelligent AI"—machine intelligence more cognitively capable than humans in all practically relevant domains—would rank among the most important transitions in history. Superintelligent machines could produce great benefits for human health and longevity, scientific understanding, entertainment, space exploration and in many other areas. Taken together, these applications would enable vast improvements in human welfare.[39]

In a different essay he further reminds us that the advent of meaningful AI brings with it a host of challenges:

> At the same time, the development of superintelligence will be associated with significant challenges, likely including novel security concerns, labour market dislocations, and a potential for exacerbated inequality. It may even involve, in the form of accident or misuse, significant existential risk.[40]

Demis Hassabis, the co-founder of one of the most advanced AI technology companies in the world (DeepMind, recently acquired by Google), noted in a recent Bloomberg article:

> I think ultimately the control of this technology should belong to the world, and we need to think about how that's done. Certainly, I think the benefits of it should accrue to everyone. Again, there are some very tricky questions there and difficult things to go through, but certainly that's our belief of where things should go.[41]

These leaders of the new revolution in machine intelligence are raising the red flag (or, at minimum, a yellow flag). Some of us paying attention; most of us are not. The tip of this spear certainly poked me during the past year.

One of the reasons for the possible indifference is that there currently is no consensus on the likelihood and timeline of the development of superintelligent AI. In opinion surveys of our experts, a majority place a significant chance of high-level machine intelligence (HLMI) being developed in the coming decades. When HLMI is defined as a machine intelligence capable of carrying out "most human professions at least as well as a typical human," the median view among a sample of the top 100 most cited AI researchers was a 10% chance of such AI being developed by 2024, and a 50% chance of it being developed by 2050.[42]

When HLMI was defined as "unaided machines [accomplishing] every task better and more cheaply than human workers," a majority of the sample of authors at two leading technical machine learning conferences ... believed there was at least a 10% chance of it being developed by 2031.[43]

In view of how much could be at stake, even a modest chance of advanced general AI being developed in the next several decades would provide sufficient reason to give this topic careful examination.

The risks are real. Two in particular are well documented in Nick Bostrom's seminal work, *Superintelligence: Paths, dangers, strategies*. In the book (which if you have the time and horsepower is worth a read or listen), Bostrom describes the first as the risk of *speed*. He imagines the scenario where separate developers of advanced AI systems skip steps to be first. The prize of being first is so high in our winner-take-all environment that the science of building test-hardened software is ignored or greatly circumvented. The cautious, thoughtful engineering team is beaten to the punch by the competitor operating on fundamentally different principles. The power to build an uncontrollable system is now in the hands of these engineering teams. Who are their bosses?

The second way is more direct: Technology from sufficiently advanced AI is purpose-built as a weapon. As Bostrom summarizes:

> *Another way in which AI-related coordination problems could produce catastrophic outcomes is if advanced AI makes it possible to construct some technology that makes it easy to destroy the world, say a "doomsday device" (maybe using some futuristic form of biotechnology or nanotechnology) that is cheap to build and whose activation would cause unacceptable devastation, or a*

weapon system that gives offense a strong enough dominance over defense to create an overwhelming first-strike advantage.

There could also be various regulatory "races to the bottom" in the use of AI that would make failures of global coordination unacceptably costly.[44]

Bostrom and his collaborators sum up the human dilemma perfectly in his oft-cited quote:

Before the prospect of an intelligence explosion, we humans are like small children playing with a bomb. Such is the mismatch between the power of our plaything and the immaturity of our conduct. Superintelligence is a challenge for which we are not ready now and will not be ready for a long time. We have little idea when the detonation will occur, though if we hold the device to our ear we can hear a faint ticking sound…

And we can't just shoo all the kids away from the bomb—there are too many large and small parties working on it, and because many techniques to build innovative AI systems don't require a large amount of capital, development can take place in the nooks and crannies of society, unmonitored. There's also no way to gauge what's happening, because many of the parties working on it—sneaky governments, black market or terrorist organizations, stealth tech companies …—will want to keep developments a secret from their competitors.

[W]hen you're sprinting as fast as you can, there's not much time to stop and ponder the dangers. On the contrary, what they're probably doing is programming their early systems with a very simple, reductionist goal—like writing a simple note with a pen on paper—to just "get

the AI to work." Down the road, once they've figured out how to build a strong level of intelligence in a computer, they figure they can always go back and revise the goal with safety in mind.[45]

The Future of Life Institute has created an "Open Letter on Artificial Intelligence"[46] that has been signed by some the most prominent thinkers, academics, business leaders and citizens of our day. A key paragraph states:

We recommend expanded research aimed at ensuring that increasingly capable AI systems are robust and beneficial: our AI systems must do what we want them to do.... This research is by necessity interdisciplinary, because it involves both society and AI. It ranges from economics, law and philosophy to computer security, formal methods and, of course, various branches of AI itself.

In summary, we believe that research on how to make AI systems robust and beneficial is both important and timely, and that there are concrete research directions that can be pursued today.

The signatories of this "Open Letter" cut across the spectrum of science, theology, policy and the business community. Perhaps man is learning about this second law of disruption and getting ahead of its existential threats? The "Open Letter" is, in my opinion, an underappreciated effort to share the best of the world's leading thinking on a technology that we are only just coming to grips with. These signatories are looking squarely at the question: "Just because we can, does it mean we should?" and forcing us to think about scenarios far ahead of what is actually possible. In my research it is also apparent that many of the signatories implicitly understand the three laws of disruption:

- AI progress and capabilities, especially once AI algorithms begin creating new AIs, will come at us exponentially,

- Many of their uses will be unknown, possibly unpredictable and not going to disappear once invented, and

- Our current forms of regulation, and regulatory frameworks and organizations, are inadequate.

We will address the AI discussion again at the end of the book as we look at the existential crises facing the planet and what we can do to head them off.

The Latest and Last—Playing God: *Editing the Human Genome*

On July 31, 2017, as I was writing the final parts of this book, my research feed kicked up the news that U.S. scientists had used CRISPR (Clustered Regularly Interspaced Short Palindromic Repeats) technology to genetically alter human embryos. There had been a two-year international moratorium in place on the technology because of the controversy surrounding it. Authorities in the U.S. have called CRISPR "a threat to national safety, citing the ease of use and rapid rise of the technology that enables scientists to snip out any fragments of DNA they wish by programming an enzyme that acts as a sort of scissors." Through the CRISPR process, scientists can perform germline editing, "permanently changing the genetic makeup of an individual and all that individual's offspring for generations to come as they would no longer pass on the diseased gene."[47]

In CRISPR we have a technology, following the curves of Moore's Law we discussed previously, that is bringing the alteration of living system's genetics out of the hands of the extraordinary few to the many. A new genie emerges.

As mentioned, scientists (including a CRISPR co-inventor) have urged a worldwide moratorium on applying CRISPR to the human germline, especially for clinical use, stating that "scientists should avoid even attempting, in lax jurisdictions, germline genome modification for clinical application in humans" until the full implications "are discussed among scientific and governmental organizations".[48] In the field of Bioethics, the debate is everywhere. This clearly is a genie of human invention that deserves and requires significant thought and oversight. The science community is in agreement on this point.

As we know, the debate over genetic manipulation has been ongoing for many decades, with the first regulation of the technology happening in the early 70s when the first uses of Recombinant DNA were commercialized and the regulatory framework in the US and other countries needed to catch up. The genetically modify organisms (GMOs), others have argued, are the primary reason we can feed seven billion people today.

But as the CRISPR discussion reminds us, it's one thing to have pest-resistant corn as a result of fine-tuning the genes of the plant and an entirely different thing to be able to alter the genome of the human species.

In a February 9, 2016, worldwide threat assessment report to the Senate Armed Services Committee, James Clapper, the U.S. Director of National Intelligence, added gene editing to a list of threats posed by "weapons of mass destruction and proliferation." As the *MIT Technology Review* noted on that day,

Gene editing refers to several novel ways to alter the DNA inside living cells. The most popular method, CRISPR, has been revolutionizing scientific research, leading to novel animals and crops, and is likely to power a new generation of gene treatments for serious diseases. It is gene editing's relative ease of use that worries the U.S. intelligence community, according to the assessment." Clapper's report said: "Given the broad distribution, low cost, and accelerated pace of development of this dual-use technology, its deliberate or unintentional misuse might lead to far-reaching economic and national security implications."[49]

But the fact is the genie of gene editing is out of the bottle. Under regulatory frameworks, we likely will have no concerns; however, there are major concerns that the technology's decreased cost and increased availability will result in its finding its way into for-profit commercial entities and (worse) failed nation states.

A research paper first published in *Science* in March 2015 hits this nail square on the head. "A prudent path forward for genomic engineering and germline gene modification" concludes with the recommendation that the use of CRISPR/Cas9 and subsequent technologies follow four principles:

1. The strong discouragement of any attempts at germline genome modification for clinical application in humans
2. Open forums for experts to share information transparently about the risks and opportunities
3. Transparent research
4. Creation of a globally representative group to "further consider these important issues, and where appropriate, recommend policies"[50]

This reaction and approach is admirable, thoughtful and coherent. But we need to recognize that the traditional organizations that deal with matters of such complexity tend to operate at a glacial pace. The United Nations and other global agencies all have the right ethos and spirit. Can they possibly operate at a pace that keeps up with the exponential and combinatorial technologies?

Do we need something different at the core of these well-meaning bodies that propels the strength of their principles with a new velocity founded on a social contract of trust?

Summary

Throughout history, mankind's attempts to manage the "genie in the bottle" have often lead to unintended consequences. The purges of the Dark Ages led to the Reformation and Renaissance, the rise of democracy and the modern education system. Great leaps forward in technology have even been used as explanations for the fall of ancient civilization: "What happened? Where did the culture go?" But the best explanation for me has always been more akin to that of the nerdy scientist from the first *Jurassic Park* movie (the quirky and brilliant Jeff Goldblum) when he's describing the reproducing Tyrannosaurus Rex, "Nature always finds a way."

True, too, of the human species. Humans often find the way. But it's not by putting technologies back in the bottle.

As we now know, the acceleration is something unheard of in our history. If we look at the speed of change as measured by the time frame between existential changes, we get something that looks like this:

THE EXPONENTIAL TIMELINE OF EXISTENTIAL TECHNOLOGY ADVANCES

Time Frame (power of 10)	Existential Technology Transition
10^6	Fire to Stone Tools
10^5	Stone Tools to Wheel
10^4	Wheel to Printing Press
10^3	Printing Press to Electricity
10^2	Electricity to Flight, Computing
10^1	Computing to Thermo-Nuclear Capability, Climate Change, Human Gene Editing, Singularity, Inter-planetary Travel

In putting this table together, I realized that we are on a rapid countdown to 10^0! The transitions that have previously taken thousands of years are now coming at us in single-digit years. If Elon Musk and others have their way, we will become multi-planet species by the time Olivia and Mervis are 20.[51]

This compression of time deeply impacts Law #2. Simply stated, the technology genie going back into the bottle is impossible and I suggest there are three reasons this is so:

1. Not only is the slope of the curve increasing at an increasing rate, the availability of information on the subject matter is openly available and widely distributable. The open-source nature of the Internet results in the **Velocity** vector—technology and invention spreads fast.
2. As the speed increases, we have a corresponding decline of global leadership, the fall of the nation state and the decline of online discourse. This is the **Tone** vector.

3. The Law of Permutations resulting from multiple, exponentially changing technologies, many of which are combining, results in the **Volume** vector.

This means that the **spread** of the technology genie is also impossible to manage. The combination of these Speed, Tone and Volume vectors solidifies the fundamental impossibility of rewinding technological discoveries like never before. To paraphrase my *Jurassic Park's* Mr. Goldbloom, "If Nature always finds a way, the real question is: Can mankind find a way?"

Sometimes I feel like we are here:[52]

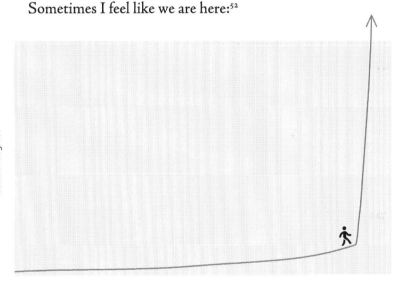

If the rate of technology advance is increasing and we can't put it back even if we want to, then what are we doing about it as a species? This is where Law #3 comes at us. And, in case you want to stop here, the conclusion is simple: we're not ready.

chapter 4

LAW #3

OUR LINEAR SYSTEMS OF ORGANIZATION

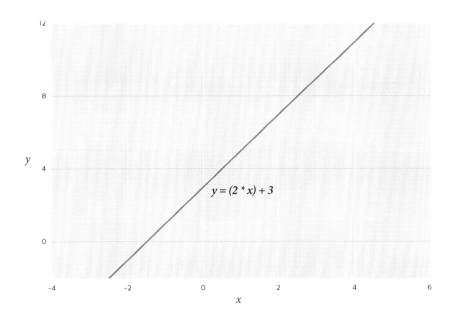

Almost all systems of our world evolve linearly. Human systems of organization are no exception. By any measure—whether expressed as the amount accomplished in a given time period, or the ability to adapt to an external change as expressed in time—most human organizations that bring together disparate groups of people move slowly and predictably.

As a quick math reminder, linear growth is expressed as an algebraic expression. In the linear graph example above, the function is y = (2*x) + 3. This is more accurately called "arithmetic" growth. It is in contrast to our friend "exponential" growth (for example $y = ab^x$).

Basic math, right? But the important thing to remember is that almost all natural and man-made "systems" of organization follow a similar change profile. Over time—whatever the "measure" of change is—all systems evolve, grow and shrink at a fairly regular pace. Seen over smaller time frames these changes can appear exponential, happening over leaps, or as a much "steeper" line, but taken over a long enough period of time, systemic change is primarily linear.

This principle of system change governs such things as:

- The evolution of our solar system

- Human development

- Evolution of species

- Laws and legal frameworks

- The professions

- Democratic systems of governance

- Education curricula

- Religious doctrine

- Human performance capabilities (athletics, for example)

The challenge with linear systems is that when faced with large change events, they often struggle to react. With sustained exponential change, the outcome is far less certain. In fact, short of some of the catastrophic events of the first two World Wars, we have no precedent where our most fundamental organizational systems have been faced with sustained exponential change all at the same time. And that is a problem since the first place we often look for guidance are these very organizations.

Let's look at a one of these in some detail to understand the problem and where we stand.

Government

A clear example of how the tip of our proverbial spear has moved very far from the tail is found in how our systems of government are simply not able to keep pace with the changing nature of the world of technology and the speed and importance of innovation. The populace is increasingly finding itself at odds with personal technologies and work systems vastly more efficient and effective than those of government, including the ability to participate in the decision-making and innovation processes of the governing bodies. As former U.S. President Obama recently said we're the only advanced democracy in the world that makes it harder for people to vote than to order a pizza. How do we redesign our systems so we don't have 50% voter participation?

Worse, the increasing distrust and cynicism of the developed and the developing world of our governance systems is at an all-time high. My previous discussion of Mr. Trump is the perfect example. And it is a frightening reality.

Of all of the innovation and responses to change I have seen and researched, it is the ability to re-shape how our governments function in a connected world that has the potential for the largest impact on closing disconnection inherent in modern democracies—the inequality gap. This is where the tail meets the 21st century.

How do we do this? Is it simply a matter of investing our governments with the same tools that our most innovative private companies use to innovate.

Not so simple, as it turns out.

The challenge of bringing technology into government is no simple panacea. From an essay by the World Economic Forum in May of 2016:

> *Part of the explanation is that technology is a problem as well as a solution. "There's a lot of technological triumphalism about how [the Internet] can be used to improve democracy," says the columnist and former White House speechwriter David Frum. "But in the end what seems to have happened is that it's empowered angry and highly motivated minorities, and empowered them to slow down the system. Getting things done seems to go slower and slower every decade. How long does it take to build a highway? How long does it take to build a bridge? How long does it take to get a presidential nominee through the Senate?"*[53]

The result is what Francis Fukuyama has labelled "vetocracy," in which it is much easier to stop things than start them. Even the admirable devotion to increasing official transparency has sometimes been counter-productive, "like creating a big Amazon rating system for government that only allows one- or two-star ratings,"[54] according to Archon Fung of the Kennedy School of Government.

Robert Corville's excellent chapter on the effectiveness of Government in his book *The Great Acceleration: How the World is Getting Faster,* summarizes thusly:

Another problem, which it is impossible to overestimate the extent of, is the pressure of the accelerated news agenda. Put simply, those in government often lack the time to think—to take time to chew through problems and come up with policies rather than being forced to respond to the latest gyrations of the 24-hour media cycle.[55]

However, even with these challenges combined with added security, more trust and an algorithmically driven, automated future now in the cards, governments continue to look to the Internet to transform public services. So as well as regulating it as a utility, governments need to innovate around the Internet – and that's something the commercial world is dragging its heels on.

The "Fourth Industrial Revolution"—as the World Economic Forum calls it—is real[56]. From their seminal work in 2016 they note—on the role of Government,

As the physical, digital, and biological worlds continue to converge, new technologies and platforms will increasingly enable citizens to engage with governments, voice their opinions, coordinate their efforts, and even circumvent the supervision of public authorities. Simultaneously, governments will gain new technological powers to increase their control over populations, based on pervasive surveillance systems and the ability to control digital infrastructure. On the whole, however, governments will increasingly face pressure to change their current approach to public engagement and policymaking, as their central role of conducting policy diminishes owing to new sources of competition and the redistribution and decentralization of power that new technologies make possible.[57]

The ability of linear systems like government and public agencies to understand the need for change and—more importantly—find the will and capacity to adapt, will seal their fate. As the WEF notes, however, transparency and efficiency are particularly hard for these organizations.

But here is a more systemic issue: Over the past 30 years or so, starting primarily with the conservative policies of the Reagan/Thatcher era in the western world, we have successfully harangued government to be less. Smaller is better. Less is more. The problem with that is systemic, dramatic change to the way government works requires large scale investments in people, processes and technologies over time. No politician in our modern era would dare campaign on the slogan, "Let's spend more now so we can be much smarter and spend much less later!"

To think outside of the box and create exponentially smarter institutions that use less resources requires a new kind of non-linear thinking. Here is a suggestion: The next time you meet with someone who has even a small amount of control over your life –for example, your local, regional or federal politician, or your kids' school teachers or principal— be prepared to ask a simple question: "What does the future hold for you?" Ask them to describe what their world will look like in five years.

It is a simple question but the answers are shockingly revealing. I recently asked a politician what he understood as the cause, effect and implications of Uber for the city. (Our city recently and surprisingly rejected Uber.) His answer was naïve at best and economically harmful to the city at worst. His answer centered on providing protection to the jobs and investment already made by the taxi industry. I said to him, with respect, that the ability to answer that question told me immediately whether he understood the digital economy (and frankly whether I would spend any time going forward with him or his campaign). Uber represents so much more than the disruption of the taxi business. It is asking us to rethink our modes of transportation and to reimagine neighbourhoods originally designed to encourage the prevalence of two-car garages and the associated suburban architectural wastelands resulting from these garages. Driverless, electric cars are coming; Uber is not about disrupting the taxi business!

Now, before you condemn me as a digital elitist, think about the policy implications of such a question. It goes to the heart of the role that the government plays in establishing the rules of the game, and the speed at which they do it. It is the speed and depth of any ruling party's response that is under siege at the moment. Given enough time, any political organization (with perhaps the exception of the US Congress at the moment) can figure out a civilian-appropriate response.

But things are moving far too quickly for traditional responses. Public policy development and decision-making in many of our institutions is based on what Klaus Schwab calls Second Industrial Revolution thinking which designed processes to be linear and mechanistic. We see it all over the world—and not just in the usual suspects of Government agencies and other public bureaucracies but in some of our largest businesses. These organization are being hit head-on by disruption and are proving unable to cope.

Schwab gives us a hint as to where this is headed:

> *How, then, can they preserve the interest of the consumers and the public at large while continuing to support innovation and technological development? By embracing "agile" governance, just as the private sector has increasingly adopted agile responses to software development*

and business operations more generally. This means regulators must continuously adapt to a new, fast-changing environment, reinventing themselves so they can truly understand what it is they are regulating. To do so, governments and regulatory agencies will need to collaborate closely with business and civil society.[58]

The last sentence makes a fundamental point. This new social contract is essential for establishing a new way of engaging with our "linear" government systems. More to follow on that subject!

Other Linear Systems

As I noted at the beginning of the chapter, there are many examples of linear systems under systemic pressure due to advancing and exponentially changing technology. The global education system is under incredible stress—from the four-year "degree factories" that many post-secondary institutions have become, to an increasingly underfunded public education system in the developed economies to the massive gender inequalities found in the developing world. Organizations like the World Economic Forum have called for a deep and immediate revamp of the skills that are needed in the digital economy, but what is often missed is the recognition that the wheels of these institutions are slow to turn. They are structured to evolve through peer review and careful considerations. The organizations themselves, not just what they produce, have to be changed.

Education in the Developed World

In the developed world, the education playing field has a very different set of challenges; the white-hot technologies at the tip of our spear and future ways of work are completely out of sync with the way we are educating our children. Even worse, an increasing amount of data supports the idea that higher education in certain areas of study does not guarantee success.

Sal Khan, the founder of the extraordinarily successful Khan Academy, which is providing online knowledge, skills and education in ways never before seen, and at a marginal cost (approaching zero) that invites universal access, says in his ground-breaking 2014 TED talk 2014 speech that the "pyramid of education and training is inverted"[59] and he seeks to turn it upside-down. The core skills that traditionally made up primary and secondary education need to be automated, transforming teachers into creative and design thinkers—changing the value of the teacher as well as how individuals acquire knowledge and skills throughout their lifetime. It is a powerful argument for the need to change and an acknowledgement that we have at our fingertips the tools we need to deliver knowledge in new and innovative ways.

Perhaps the greatest opportunity of all is the ability of the developing nations—supported and facilitated by institutions and organizations of the developed economies—to leapfrog the largely "analog" education experience of the 20th century and take full advantage of the tools, connectivity and insight we have acquired in the past 20 years.

At a more foundational level, the exponential march of technology and the seriousness and impact of the rapid, unintended spread of new innovations call into question the ability of our democratic/capitalist society to function. That is a treatise for an entirely separate book. I have included some of the best of the reading and research on this subject on this book's website. *thespear.co/research.*

Summary: *The tip of the spear is white hot*

An exponentially exploding set of existential threats and opportunities that can never go back in the bottle are being met by biological, ecological and human systems of governance that move at a pace of their own making.

The tip of the spear is hot. In fact, it is as hot as it will ever get for our species. Moving forward, we will look into the evidence that the cracks are beginning to show. A simple review of the discussion, research, literature and best of human thinking sees the following signs that are a cause for deep concern:

- Global naiveté about the coming changes

- Lowest common denominator thinking

- The decline of community and discourse

- The decline of the nation state/death of democracy

- An education system ill-prepared for the next generation

In short, in the face of a many-sided existential crisis, the human species has to decide whether it will behave more like its distant and defunct evolutionary relatives or exponentially rise to the challenge and evolve at a pace that it never before has done. The race is on.

PART 2

MEET THE TAIL

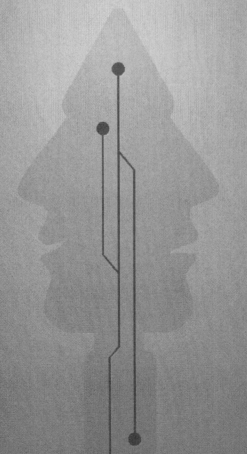

chapter 5

INEQUALITY
THE TIP MOVES FAR AHEAD OF THE TAIL

The race is on. To determine the state of that race, let's remember my ah-ha moment about Donald Trump earlier in the introduction and look a little closer.

He is the butt of the spear isn't he? Or perhaps he represents the rabble outside the binary walls of the digital kingdom. Trump is simply an amplifier for a growing class of human beings who are increasingly at odds with the world around them. He is the manifestation (or more likely the exploiter) of disenfranchised and angry humans who sense (and some rightly so) they are being left behind and are "mad as hell" and "not going to take it anymore."

I also believe he represents a hint as to what a dystopian future might look like. I am concerned because history has shown many times that, when the "ruling" class gets too far ahead of the masses, heads roll—literally, in many cases. As noted, we have witnessed incredible technological change in the past 10, 20 and 50 years but I can assure you that the "reptilian brain" that drives our basest needs and desires has changed very, very little in the same time period and some would argue that on the (linear) scale of evolutionary time frame, there isn't a chance it could.

As we saw in the previous chapter, even more troubling is that the very systems of governance, market forces, laws and other fundamental frameworks are simply incapable of moving at the speed of change. Not even close. Think of the standoff between Apple and the FBI, the dilemma Netflix poses to the CRTC or FTC, or the havoc Uber has wreaked on the taxi industry. The free market alone will not "solve" this calculus of change.

On the other side of the equation are the makers, inventors and those profiting from this digital disruption. My hard glare at them is simple. Ask the fundamental questions: *Just because I can, does it mean I should? And if I should, then how?*

In this coming age, the fundamental laws of economics are being transformed; the spoils will go to those who adapt the fastest, those who "create" this new future and all of its component pieces. Like the exponential technology trends driving this change, human adaptation will begin to create **exponential** inequality. Left unchecked, the rewards that will return to those who can **exponentially** adapt will be staggering. Unlike the robber barons of the 19th century and the technology titans of the 20th century, the lead actors and winners will become larger than the nation states that are desperately trying to keep up with the governance necessary to manage the implications of this new concentration of wealth.

We are seeing signs of this new era, aren't we? Nothing exemplifies this more than the history of the leading organizations from the 90s and those of today. Back in 1990, the largest employers in the North American economy included the big three automakers—GM, Ford, Chrysler. When we compare them to the big three tech companies of today, the economic realities are startling. In 1990, the trio of American automakers together brought in $36 billion in revenue altogether, and employed over one million workers; Apple, Facebook and Google, which together bring in more than $1 trillion dollars in revenue, employee only 137,000 workers.

The top four companies in the world in terms of market capital are all technology companies. Their combined value as of this writing is over $2.5 trillion dollars. Only the United States, China, Japan and Germany are larger in terms of GDP. Their cash balances, of almost a half a trillion dollars, are larger than all but three countries' total foreign currency reserves.

In an article called "New World Order: Labor, Capital, and Ideas in the Power Law Economy" published in July 2014 in *Foreign Affairs*, Eric Brynjolfsson, Andrew McAfee, and Michael Spence (a Nobel laureate and professor at New York University) argued that "superstar-based technical change … is upending the global economy." That economy, they conclude, "will increasingly be dominated by members of the small elite that innovate and create."[60] If technology leads to more inequality, it may have the effect of suffocating demand from the economy and can become self-destroying, making further technological evolution redundant as only the few could afford it. The authors conclude:

> *The growing inequality around the world can no longer be ignored, and addressing this and the other problems of capitalism, such as environmental degradation, is not only the morally right thing to do, but the pragmatic thing to do.*[61]

But these are just symptoms of the real problem. And it is what others have called *the* existential issue of the day. It is truly a "wicked problem."

The wicked problem

It has been called many things but, simply stated, the Tail is really just another way to describe inequality. If you thought "disruption" was a loaded word, try discussing inequality across political, socioeconomic, racial, geographical and gender lines. It is a charged, single issue that unites us in our differences. And it is about to get a lot worse.

To start, here is the factual evidence of the problem in a single chart:

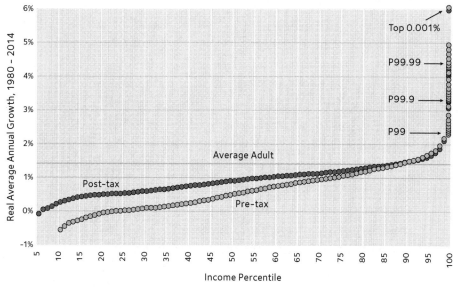

gabriel-zucman.eu/usdina/

I have many, many charts exactly like this one. In every case—no matter the perspective, time-scale or data source—they point with shocking clarity to the facts of inequality. The rich are getting richer. The current tax reform debate underway in the US brings the issue directly to the pocketbooks of US citizens and voters. The developing world fares no better.

If you look at the data, the facts are staggering. In North

America, the middle class pay raises prior to 1980 were about 2% per annum on an inflation adjusted basis. This translates into about a doubling every 34 years and is, by any economic measure, a substantive rise in the standard of living of the largest group. These rates of increase were in fact *higher* than that of the rich.

After 1980, things change dramatically and the gap between the super-rich and the rest of the large middle class began to increase *dramatically*. The super-rich—those in the 1/40th percentile are now the only group seeing these types of increases.

David Leonhardt in the *New York Times* (Op Ed, August 7, 2017) summarize the implications of this data simply:

The basic problem is that most families used to receive something approaching their fair share of economic growth, and they don't anymore.

This is a profound problem of our time. But perhaps we can better understand inequality through the long lens of human history and our "evolution" and adaptations.

The initial reason the human species came to dominate this planet was our ability to intelligently adapt to a changing physical environment. Once we had begun to fill the "cognitive void" in the evolutionary ecosystem, we somehow had to figure out how to live with ice ages, heat, natural disasters, food production and transportation to combat all that the natural environment threw at us. We managed to do so because of our cognitive ability, and we survived—but barely.

Inequality in these earliest human epochs was represented by the capacity and development of the frontal lobe and other brain substructures that emerged and evolved in time frames measured in the hundreds of thousands of years. The *disruption* as a result of this inequality was that natural/evolutionary causes killed off several promising but ultimately failed limbs of the human evolutionary tree. *Homo sapiens* emerged and the disrupted species died off.

The second stage in our evolution—one that we are still in—has been our ability to control and alter our environment. It is presented in our ability to use technology to increase our standard of living and well-being and to support an exponential increase in population. *Inequality* has been caused by an unequal access to natural resources and the rise of capitalism—the invisible hand of the "market" and the misguided presumption that this effect was benign and purely efficient—devoid of anything that reflects the irrationality of man. The timeframe of this epoch has been thousands of years. The central *disruption* is the elimination of civilizations due to war, poverty and the rise of a continually changing guard of globally powerful nation states.

From all signs, we are about to enter the final stage of our evolution, and it will indeed be exponential in both effect and timing. It will be our ability to create the environment of our choosing, be it on this planet or others. We will be able to manufacture and "update" our species and others at will, manufacture/terraform new worlds, create new intelligence. *Inequality* will be driven by access to the tools of adaptability. *Disruption* will merge as multiple existential threats to our species (and to many, many others) the survivors of which will be a very select few. This will be the final expression of inequality as our species makes its final "all-in" bets.

Now, before I get my libertarian and Ayn Rand followers frothing, I am simply pointing out the obvious. I am all for the survival of the fittest when the rules of the game are known and the players are equally distributed, but I can assure you as I have in earlier paragraphs, when the tip moves too far from the butt, revolutions happen. And that distance has a word: Inequality, and the reason for it is an unlevel playing field.

This is a conversation from eternity, isn't it? The world's game has been played on an unequal playing field since the beginning of time.

But here is the point:

For the first time in our human history we have the tools and capability to create the conditions for equality within a generation.

The first time. Ever.

There has never been such a potential benevolent tipping point before in human history. Something's knocking at the clubhouse door and it's either opportunity or the steel-toed boot of the mob.

This is very much our choice—and I am painting this in a black and white "this OR that" way purposefully because I believe the stakes are that high. To repeat, the tip is speeding away from the tail, and if left unchecked, our world is going to look very different within this generation. As Stephen Hawking wrote in *On Inequality*:

> *If machines produce everything we need, the outcome will depend on how things are distributed. Everyone can enjoy a life of luxurious leisure if the machine-produced wealth is shared, or most people can end up miserably poor if the machine-owners successfully lobby against wealth redistribution. So far, the trend seems to be toward the second option, with technology driving ever-increasing inequality.[62]*

Unilever's CEO, Paul Polman, put it more succinctly, "Power is dispersed, but wealth is concentrated."[63]

FREEDOM "FROM" VERSUS FREEDOM "TO": A NEW INEQUALITY

The subject matter of this book is—to no one's surprise—very fluid. Perhaps the biggest challenge has been to filter through the ever-changing inputs and concepts and lay down opinions and thoughts that have both resonated with my research and 30 years of thinking yet are clear new patterns.

One of these new patterns has been the idea that the very presence of a global infrastructure for communication has created both the means for delivering sometimes devastating change in our societies but is unequal in its ability to heal and re-connect ideas and societies. This is powerful new thinking and—when coupled with our three laws of disruption—creates barriers and challenges to creating systemic change and the cultural evolution of our species. It is a stop light (or a solid merge sign) at the crossroads.

In the excellent treatise on technology and inequality called *Thank You for Being Late: An Optimist's Guide to Thriving in the Age of Accelerations*, Thomas L. Friedman—of *The World is Flat* fame—makes some razor-sharp observations on both the power and importance of technology in making change happen. This is especially true as he discusses the notion of technology's ability to inspire freedom "from" something versus freedom "to" a new condition.

> *There is mounting evidence that social networks make it much easier to go from imposed order to revolution than to go from revolution to some kind of new sustainable, consensual order.*[64]

This is a shocking and unintended consequence of the new found power of a connected society, isn't it? I believe we all thought that we had the perfect tools to spread, for example, the news of where and when a protest was to take place or the online posting of new manifesto demanding change. And we believed those same tools could be used to re-connect and reposition ideas, societies and political agenda. Then the messy culture of human nature kicked in. We have learned that just because we can now tear something down more quickly because we spread emotion, fear, information and dis-information, the crafting of new solutions requires much more subtle and complex human behavior—most of which rely on a culture and foundation of trust—something our cold technology is unable to do.

This freedom "from" has been a gift of the gods; the freedom "to" is something else. As Dov Seidman, CEO of LRN, which advises global businesses on ethics and leadership, notes in Friedman's book:

> "'Freedom from' happens quickly, violently, and dramatically"... "'Freedom to' takes time. After the Jews got their freedom from Pharaoh in Egypt, they had to wander in the desert for forty years before they developed the laws and moral codes that gave them their freedom to."[65]

Elsewhere, the difficulty of achieving genuine political order has led to growing numbers of "un-free" people in the world. Income inequality is destabilizing, "but so is freedom inequality," says Seidman. When "freedom from" outstrips "freedom to," amplified actors in the grip of destructive ideas "will cause more harm and destruction, unless they become inspired and enlisted in constructive human endeavors," he argued. "They will be like inmates on the loose."[66]

Wael Ghonim, aka "the Google guy," who helped launch the 2011 revolution against the Egyptian president Hosni Mubarak with a single Facebook page, said in 2016:

> *Five years ago, I said, "If you want to liberate society, all you need is the Internet." Today, I believe if we want to liberate society, we first need to liberate the Internet.*[67]

This is a staggering indictment of the very technology that we had pinned so much hope on. In five years! Unresolved, the hijacking of the most powerful tool ever created (thus far) by the human species within a decade, by the worst of what makes us human is a sure sign we are not ready to evolve.

Finally, Harvard University strategist, Graham Allison says,

> *Historically, there has always been a gap between people's individual anger and what they could do with their anger. But thanks to modern technology, and the willingness of people to commit suicide, really angry individuals can now kill millions of people if they can get the right materials.*[68]

And if we apply our three laws of disruption, is becoming steadily easier with the globalization of flows and the rise of 3D printing, by which you can build almost anything in your basement, if it can fit.

This new inequality—expressed here by some of the best thinkers of our time—lies on top of our baser needs as a species.

It is both surprising and concerning.

chapter 6

INEQUALITY

CHALLENGES IN THE DEVELOPING WORLD

So far we have spent most of the book with a developed and western economic viewpoint. As we have seen, that view holds sufficiently scary propositions, filled with major challenges caused by the three laws of disruption—the exponential technologies, spreading increasingly unregulated into a world ill-prepared to cope.

Let's move our focus to areas of the world that have an entirely different set of challenges.

Africa, Democracy and Change

As we are reminded by anyone with even a passing overview of the continent, Africa is not a single unit. It is a complex and multi-faceted, multi-jurisdictional region with a crushing colonial past that presents its leaders and the world with an extraordinary range of challenges and opportunities.

At the highest level of thinking about technology disruption, change and the tip of our spear, Africa is probably the area of the world where the most disruptive changes will occur over the medium term. Ironically, the continent's lack of infrastructure may prove to be one of its greatest assets. It has forced entrepreneurs to contrive different solutions and jury-rig the limited resources available—primarily the mobile phone. Africa has already taken the lead when it comes to mobile banking and mobile health, and implemented those solutions more quickly than anywhere else. There's a burgeoning sense that an app like Apple Pay wouldn't be deemed newsworthy, a frog that's leapt already.

Some data that we examined previously:

- 700 million people will be moving into African cities in the next 35 years—that means an entire New York City has to be built every six months until 2050.

- Nine out of the 20 fastest growing economies in the world are in Africa.

- African startups raised $187.5 (USD) million in 2016.

- African venture capital companies reported up to 330% growth last year.

- Africa is home to 1.2 billion people and 200 millions of these are aged 15–24

- Africa has the fastest growing middle class in the world.

- Smartphone usage is at a tipping point, expected to reach 400 million users in 2020.

As new investment and capital begin to flow into the

African continent over the next several decades, a natural outcome will be the harsh glare of the capitalistic due diligence spotlight. While early entrants into distressed regions often represent the worst of the self-interest businesses, history has shown that over time, capital flows from transparent nation states and global corporations tend—albeit imperfectly—to norm (i.e., create effective oversight, governance and a reasonable return to capital). As we have found in all things in the *Tip of the Spear* world, scrutiny and iteration are often brutal but largely effective.

In short, turning the flywheel on the sub-Saharan African continent—for example—is likely going to start with capital and investment. What's new this time is that transparency and improved technology availability and literacy will move change along much more quickly. Corrupt regimes that fail to provide transparency will fall with alarming speed and, ideally, with less intensity than the sectarian chaos of the so-called Arab Spring.

In summary, global governance inequality is at the core of the tip of the spear problem. In the developed world, we are seeing the arrival of new technologies that are providing the potential for quantum changes in the control and access OF the citizenry and BY the citizenry. In the developing world, especially in Africa, capital inflows and entrepreneurial ecosystems are leading the way in turning the spotlight on inept governance and the benefits of a proper education

Inequality: *The Unbanked*

Safe, efficient access to the fundamental tool of the economic game—the bank account—is staggeringly varied across the globe. Do a gender, education or region pivot on that data set and you will quickly discover that what we absolutely take for granted in the developed world is not an easy thing. That is inequality.

Don and Alex Tapscott noted in their ground-breaking 2016 book, *The Blockchain Revolution: How the Technology Behind Bitcoin is Changing Money, Business, and the World,* the problem of inequality isn't necessarily capitalism itself.

> [F]inancial and economic exclusion is the problem. Fifteen percent of the population in OECD countries has no relationship with a financial institution, with countries like Mexico having 73 percent of the population unbanked. In the United States, 15 percent over fifteen years of age, or 37 million Americans, are unbanked.
>
> Financial inequality is an economic condition that can quickly morph into a social crisis. ...The problem is that most people never get a shot at seeing the benefits of the system because the Rube Goldberg machine of modern finance prevents many from accessing it... Blockchain will have the greatest impact in areas where the payment networks don't exist or are very poor...Blockchain will push many nascent initiatives, such as mobile-money service providers like M-Pesa in Kenya, owned by Safaricom, and microcredit outfits globally, into high gear by making them open, global, and lightning fast.[69]

Africa is arguably more suited for Blockchain and cryptocurrency adoption than many of its more developed Western peers. A primary reason is that the hard currencies of developed countries already satisfy the needs of most of their citizens, in contrast to many African currencies and payment systems, which fall well short.

A key example is remittance payments. These are a vital source of income for a number of African countries, with the average migrant supporting between 10 to 100 people, according to the World Bank. Nigeria is Africa's largest receiver of traditional wire-transfer payments from out of country residents with an estimated US$19 billion in receipts, followed closely by Egypt with US$16.5 billion.

The use of cryptocurrencies could significantly reduce, if not eliminate, the high transaction costs of remittance payments, encouraging further capital flows to African countries most in need. Africa's leading Bitcoin startup, BitPesa, has already proved this to be true. Founded in Kenya, BitPesa provides an online platform to convert digital currency such as Bitcoin into local African currencies. Its chief executive Elizabeth Rossiello, has claimed that Bitcoin-based remittance services have reduced the costs of international payments by 75%, and reduced the average time of settlement from 12 days to approximately 12 hours.[70]

Access to the Internet

From the "State of Connectivity Report 2016" commissioned by Facebook, we learn of some sobering statistics:

The developed world is largely online, but the developing world is a long way behind. Urban areas are connected; many rural areas are not. The less money you have, the less likely you are to be online. In many countries, women use the Internet far less than men. And even if the entire world lived within range of the necessary infrastructure, nearly a billion people remain illiterate or otherwise unable to benefit from online content.[71]

The Internet is a catalyst for broader social and economic advances through the access it provides to education, economic and employment opportunities, and even healthcare.

It is a critical tool for development and should be available to everyone. Acknowledging the importance of connectivity and the need to bring more people online faster, the United Nations (UN) General Assembly and the International Telecommunication Union (ITU) recently called on the international community to provide universal access to affordable Internet by 2020.

Education in the Developing World

The single greatest determinant to economic success and the demarcation line between developing and developed nations is literacy.

While the number of illiterate persons has fallen over the past decade some challenging statistics are present:[72]

- 774 million adults—64% of who are women—still lack basic reading and writing skills.

- In 2011, the global adult literacy rate was 84.1%, compared to 89.5% for youth.

- More than three-quarters of illiterate adults are found in South and West Asia and sub-Saharan Africa.

- In 21 countries fewer than 50% of the children of primary school age learn the basics in mathematics.

- In 27 countries, nine out of 10 of the poorest young women have not completed primary school.

It reminds those of us lucky enough to have the tools and skills to able to read this book that the global experience of education inequality is real and in these specific regions the tip stretches furthest from the butt. It is here the three inequalities are obviously and completely interlinked: Corrupt and impoverished governance leads to lack of investment in basic life infrastructure which in turn makes financial access impossible. It is a seemingly unending cycle to change.

But there is hope.

There are many trends that have surfaced in my discussions and research, but three stand out:

The Rise of Transparency: As we discussed in the section on leveling of the governance playing field section, the gradual disappearance of corrupt regimes combined with the dramatic cost reductions and availability of technology is enabling citizens to demand essential services such as primary education and literacy.

Leapfrogging Technology: The "exponential path" is possible as we begin to see developing countries literally by-pass entire generations of experimental technologies and infrastructure investments to emerge digitally ready. It is as though smart, well governed nations could have the ability to "disrupt" developed countries—much like the upstart firms in Clayton Christensen's "Innovator's Dilemma"[73] were able to out-maneuver the much larger and established companies because of their legacy of business and economic models.

Connected Youth: Many of the world's leaders, especially of countries that are less than transparent, open and free, need to understand that, once given access to the technology, the youth of the nation will latch hold the quickest. And once that genie is out of the bottle, the youngest generations with access to the global Internet and the rudiments of the financial systems will actually own the keys to the new economy and will, more quickly than most leaders will ever appreciate, hold the keys to the future. Leadership that resists this will not survive for long.

Summary

In this chapter, we dipped our toes into the complex subject of global inequality. There have been many excellent books, essays and symposiums on the topic. It is not a subject that I can do any justice, resolve nor even fully get my arms around in this book. But I certainly encourage the reader to do so. The axioms of *Tip of the Spear* are the "jet fuel" that is igniting what has been a growing problem for decades—especially since the early 80s. It has its roots in the revolutions of history. It is the burden of the human condition. But it must be faced head on.

In the book *Inequality: What Can Be Done?*, Anthony B. Atkinson lays out a detailed overview of the context, causes of and potential solutions for of global inequality. In his conclusion he argues (as do I) that while the role of governments from around the world is critical—in terms of taxation policy, social laws and other legislative injections—there is a new social contract in which the role of the individual and the sum of our decisions is at the centre. Atkinson persuasively argues that it is in our roles as individuals (as consumers, voters, parents, etc.) that we have the most power through our collective action and shared experiences.

Most important, he notes what we do *collectively* sends powerful signals to governments and corporations:

> Sending that email message to your elected representative makes a difference. But individuals can influence the extent of inequality in our society directly by their own actions as consumers, as savers, as investors, as workers,

or as employers. This is most evident in terms of individual philanthropy, where transfers of resources not only are valuable in themselves but also provide a powerful signal of what we should like to see done by our governments.

But, as I have stressed in the case of governments, transfers are only part of the story. Consumers make a difference by buying from suppliers who are paying a living wage, or whose products are fair trade. Individuals, acting on their own or collectively, make a difference by supporting local shops and enterprises.[74]

In short, *Tip of the Spear* is a wake-up call to the acceleration of the problems of a troubled planet. But in the timeframe during which this thesis and his book came together, the world has become an increasingly chaotic place. The three laws of disruption are happening in real time.

chapter 7

AND ONE MORE THING...

As I write this book, the summer of 2017 has provided a hint of the dystopian future that our current leaders are careening into. And I am not talking about the next decade or five years. I am talking about 2017 to 2018. Right now. This minute.

Let's take a drive through a worst-case scenario hell and remind ourselves that the world is—as many of our media pundits remind us—a very dangerous place.

> **WARNING: THE FOLLOWING GETS DEPRESSING REALLY FAST.**

There are four stages to this dangerous journey. The first is a phenomenon that is all too real and immediate to most of us who spend any time online: the Death of Discourse.

Stage 1: *The Death of Discourse*

It really has snuck up on us, hasn't it? While unruly behavior is not dependent on whether the conversation is online or offline (after all, humankind is a tribal and warring species), the conversation *tone* has changed dramatically in recent years. The causes are multiple and inter-connected.

Individuals in our human collective are suffering from a full decade or more of digital-media inspired A.D.D. Many of us have forgotten how to listen and are slowly forgetting how to critically think. We fight each other for face time, space time and air time. Worse, we allow the tools to amplify the bullies who use our gloriously connected digital medium to lower discourse to unprecedented depths of biliousness and broadcast the basest traits of our species.

In a typical encounter it doesn't start out that way, does it? Watching the reaction to the devastating fires in Fort McMurray here in the province of Alberta in 2016 (as well as other traumatic and very public events), what occurs to anyone with a modicum of humanity is that which always rises to the top—at least initially: extraordinary compassion for our fellow man/woman/child. "How can I help NOW with the basics of human needs—food, shelter, warmth?"

What then happens—inevitably and most unfortunately—is that the discourse gets hijacked. When the conversation moves online (as it always does), commentary and passive aggressive trolling pushes the discussion off the rails. Base-level human behaviours inevitably show up and things devolve astonishingly quickly. What started as a collective response to human need, spreading exponentially and positively, becomes a vitriol of the trolls.

In the case of Fort McMurray, online newspaper and Facebook discussion threads saw the deniers of climate change met head-on by the shouts of "karma" by the radical environmentalists. The poor souls who happened to be in Fort McMurray in the spring of 2016 simply needed shelter, food and water.

In the face of this, the best of our humanness disappears. "I am out of here," the best respond. Worse, it casts another shadow on our belief and faith in the collective human experience. We start to back away from the very tools that give us access to the best and the brightest.

The reasons for this are complex, but the cause and effect chain is fairly obvious: When large groups of humans remain far down the classic Maslow's "Hierarchy of Needs" for a long time, and especially when observing (often in real time) a small but entitled group who are not, they get grumpy. It takes time, but like all things exponential, it's very, very slow at first and then boom!

And here's the killer: Combining this societal bitterness with the cold, passive aggressiveness of faceless and consequence-free technology! Online trolls, bullies and thugs face no consequences and escalate things to a level that they never would in the "real" world. You don't get punched in the face for calling someone a #$#$&* in an online comment section.

And they are everywhere.

Here's a scarier thought: What will happen when the next 2 billion come online? Will they witness the utter mess of discourse and run away, or will they simply ignore and revel in their new found freedom of expression? I wonder and worry about this.

While the death of discourse is really nasty and depressing and it makes us want to stop reading crap online because we'd rather talk to our neighbours civilly over a pint (or punch them in the nose when we discover their online pseudonyms), it may frighten away the next generation of Internet users. But this is not the real problem.

Not by a long shot. But let's be clear: If we are unable to have civil conversations anymore, we are headed for serious trouble.

Welcome to Stage 2: *The Rise of Fake News*

So, we have become rather "snippy" online of late, haven't we? That is, unfortunately, the kinder twin of a deadly duo: *The Rise of Fake News and the Weaponization of Social Media.*

In all of my years involved with technology, social media and the global impact of technology, I have never witnessed the speed at which the problem of "fake news" has descended on our globally connected world.

It is not a new problem. In a recent *Atlantic* article, Adrienne LaFrance writes,

> *Fake news is everywhere. The power of the press is said to be waning. And because the nation's most famous populist—the man with his sights on the presidency—can't trust the lying media, he says, he has no option but to be a publisher himself.*
>
> *Oh yeah, and the year is 1896.*
>
> *The would-be president in question is William Jennings Bryan. In an era before the internet, television, or radio, the best way to reach the masses is with newsprint. So, without the option of tweeting his grievances after losing the election to William McKinley, what does Bryan do? He starts his own newspaper. And he uses it to rail against "fake news."*[75]

However, in the timeframe of this book's writing, I have seen the problem morph from the social media ramblings and cries of "fake news!" in the run-up to 2016 US election, to the explosion in the Fall of 2017 of a very scary potential of a global conspiracy aimed at subverting and influencing elections and democracies around the world.

Never have we had so much information at our fingertips. Whether this bounty will make us smarter and better informed or more ignorant and narrow-minded will depend on our awareness of this problem and our educational response to it. At present, I am very worried that democracy is threatened by the ease at which disinformation about civic issues is allowed to spread and flourish.

Without making too fine of a point of it, the rise of fake news has been called the new Cold War. Instead simply the threat of nuclear war and bombs (but of course we still have those, too!) we have a concerted attack by both state and non-state agents in a form of societal brain-washing that has had its genesis over the past 4 or 5 decades—primarily in the US—but certainly spreading across the globe as the Internet and social media took hold on the populace.

The decline of civil discourse (Stage 1) meeting an increasing inability to discern truth equals at best a society very much challenged to have longer running conversation (online or off) about complex issues of the day. At worst, it subverts the very heart of the democratic process. "Democracy Dies in Darkness" became the new motto on the *Washington Post*'s front page in February 2016.

It is a forewarning. Because it might get much worse.

Stage 3: *The Geopolitical Slide*

History shows us that the human species requires a good kick in the pants every once in a while. Or more accurately gives itself one. The past century saw three major wars. We ended the second one by dropping two atomic bombs that changed our view of technology and war forever. Further back as a species it appears we needed to beat the crap out of each other—religiously, militarily, ideologically—every so often. We evolve our technology and advance our species but our true nature hasn't changed very much.

We say "never again" to the bombs and to the gas chambers that created the conditions of one terrible war. What caused it? Historians will disagree on all of the exact reasons but one of the most often cited explanation is the territorial and ideological expansions of nations lead by tyrannical leaders—and I would include the United States in that category as it sought to ruthlessly exterminate the scourge of communism through the use of power and military in Vietnam and South East Asia.

Those of us born after 1970 have not had a world war to contend with, at least in the traditional sense. 9/11 changed everything, even the very nature of warfare, combat and the use of technology to deliver weapons of war, munitions as well as mass ideological influence on those most susceptible. The very same conditions that spawned ISIS/Al-Qaeda have in turn spawned an ignorant, inward-looking nationalism in the West; closed thinking and lack of education make fertile ground for ideologues and propagandists.

This cycle repeats.

And if one starts to really look closely at 2016 and 2017, we can see the potential for a daunting sequence of events that could lead to unthinkable aggression around the world or, worse, an escalation of the insidious terror from faceless and stateless terrorists. True terror from…terrorists.

But it starts innocently. It always does. We could imagine that over the next 24 months:

- Britain, fueled by anti-immigrant rhetoric and a small but very vocal majority fed up with being ignored, votes to leave the EU.

- The US, fueled by similar divides and a unique (some say might bizarre) political system exemplified by the electoral college concept, elects Donald Trump, thereby solidifying in the world's two oldest democracies at least a decade of nationalistic, 'post-fact', angry democracies.

- EU begins a natural slide into irrelevancy following Germany's decision to exit in 2018; then Italy and the rest shortly thereafter.

- The emboldened alternative-right forces find their foothold in the US and the rest of European democracies.

- China, sensing the void and lack of coherent checks and balances in the West, begins to assert itself into the vacuum left by a nationalistic US, pushes harder into Southeast Asia, dramatically raising tensions with a trigger-happy US leader.

- Pakistan and India struggling always to get along, begin to bicker—diplomatically and commercially first, then militarily in the Kashmir region. India, sensing its increased stature and importance in the region (i.e., about to become the world's largest country by population), decides to test it.

- North Korea fires another short-range missile into disputed territory while satellites confirm the testing of a long-range missiles capable of reaching the US; Trump orders a naïve and jingoistic military attack on the region.

- Russia begins running out of cash reserves as oil and natural gas prices continue to stay stubbornly low. Putin when he wins his next "election" presses aggressively into neighboring regions including China.

- Meanwhile, the protectionist trade policies of the US set off an inevitable decline in global trade. leading to massive middle class job losses.

- Meanwhile, the nascent middle class boom in the emerging economies begins to spiral down, causing significant unrest– especially in Africa—as fragile democracies and more open government falls to opportunistic dictators and strongmen fueled by the acceptance and tolerance of Trump and others.

- The Alt-Right, emboldened by their new voice online, begins to open up the cesspool of the Dark Net, paving the way for the emergence and production of new weapons of mass destruction with simple-to-use instructions and supplies.

- ISIS moves fully online now. A generation of dislocated, nation-less and disaffected Net users generates a mass cyber-attack on the global financial systems of the world, causing chaos and crippling online commerce.

- The Open Internet becomes a wasteland as brands, organizations and ultimately individuals retreat to private communication channels and the inevitable technology innovation decline begins.

Not a pretty picture, is it? Not all of these will happen nor necessarily in the order presented, but far too many can and will. The best case is that we have a pissed-off majority and a geopolitical tinderbox fueled by a transparent and increasingly negative online world. At worst, we have the beginning of our latest and likely last World War. Just when we need the best of a collective, kind human response, the best have left the building and only squatters remain.

But it gets worse.

Let's imagine the human species all woke up on the wrong side of the bed one day and then add **Gibson's Laws of Disruption** into the mix:

> # Gibson's Three Laws of Disruption
>
> **1**
> The slope of the disruption curve is dramatically increasing.
>
> **2**
> The technology genie never goes back in the bottle.
>
> **3**
> Our linear systems of human organization are unprepared for sustained exponential change.

Stage 4: *Technology Fuels a Bitter Fire*

Our venture – fueled, exponential growth-focused cabals have been spending several decades building technology for its own sake. Some technology is fueled by open source visions of greater good for all.

So Law #1 really begins to pick up speed, especially in non-information science such as biology and sensors and new devices. We start using the combinations imagined by Peter

Diamandis and others and create unimaginable technologies. We use our smart machines to create even smarter machines. We invent more stuff because it's cool.

Now comes Law #2 combined with Law #3. Because the technology genies are out and we can never put them back and because our traditional sources of checks and balances are completely incapable of managing, leading, legislating or frankly influencing, things get out of control.

Our broken capitalist system simply calls it "progress" and blindly turns its eyes because frankly it's not the nation state's problem; that is, the "market" will look after it. Remember we've just elected global "less-is-more governments."

We have now poured jet fuel on a raging fire. Because of the shrinking cost curve, we have technology-enabled a world that is now dominated by very pissed-off people with the same democratized access to all of this technology. Individuals now try to understand and react to a world gone a little crazy, friends and colleagues start turning inwards and away from others—at the dinner table, the community hall, in our cities and around the world.

A pissed-off majority, incapable of having sane, human conversation online or offline, meets a geopolitical tsunami and is fueled by technology that we are unable to control.

And guess what? We're inviting two billion more people to this party, many of whom have just risen to the first levels of subsistence, more educated than before but still with a long way to go! The slide back down their precarious and slippery slope is quick when things get ugly. Currently a billion people lack access to safe drinking water, and 2.6 billion lack access to basic sanitation. As a result, half of the world's hospitalizations are due to people drinking water contaminated with infectious agents, toxic chemicals and radiological hazards. According to the World Health Organization (WHO), just one of those infectious agents, the bacteria that cause diarrhea accounts for 4.1 percent of the global disease burden, killing 1.8 million children a year. Right now, more folks have access to a cell phone than to a toilet. In fact, the ancient Romans had better water quality than half the people alive today.

But now it is time to get a little more personal. If you really want to appreciate the urgency of this discussion, let's walk through the screenplay of the early lives of our two girls. The hypothetical journeys of Olivia and Mervis as the next dozen or so years unfold brought it home for me.

Into this near future our two girls are marching and, if it is not clear to you by now, the decisions we make in the very near term will profoundly affect the course and outcomes of their journeys

PART 3
THE JOURNEY OF TWO GIRLS

chapter 8

OLIVIA
THE GREAT SOCIAL EXPERIMENT

Remember Olivia? She's the hypothetical and all too typical 13-year-old North American teenager introduced at the beginning of the story. I could have picked many personae and many journeys through the infinite lives that exist but I chose her and her cohort because frankly her demographic has a great big bullseye on it. Not to get too far ahead of her story but the potential for change—and not the "oh-my-God-isn't-that-cool" kind of change—is staggering and potentially devastating for her. Her age group has "won" an unfortunate lottery.

Hers will be the first generation in history that will come of age under three or four existential threats, using technology no one has ever had before and with 70 plus years of lifestyle and organization structures disappearing before her eyes. All before she's 25.

Her story starts with the technology she carries with her 24/7: her supercomputer masquerading as a smartphone. I expressed my concerns about Olivia's favourite technology appendage at the beginning of the book. Over my months and years of research and with my own children and their friends, it has been quite clear to me that the only way to describe what she is about to undergo is something I have coined **The Great Social Experiment.**

Except it isn't really a proper experiment is it? No primary research, no baseline data, no hypothesis. We have essentially armed our most impressionable demographic and unfinished humans with a set of tools that are fundamentally brand new, with no idea what the short-, medium – or long-term impacts on everything from brain chemistry to social behaviours and societal interactions will be.

We simply have no historical context. None. The future is unwritten though the code exists.

I have long argued that delivering unfettered access to a global network containing every imaginable bit of human (and inhumane) knowledge delivered to devices that are constantly switched-in and switched-on to an incomplete, unfinished and highly malleable collection of young brains without discretion, oversight or fundamental research, is unfair at best and immoral at worst.

In about five years from now we will begin to see the longitudinal data from the many studies underway on the long-term impact of the continuous use of mobile phones. I am convinced that we will be looking at the single most impactful event on our youth in the history of the developed world. It will rival—and in my opinion surpass—the sending of 18-year olds to the unspeakable crucible of two successive world wars.

Let's play out the first act of Olivia's journey.

She enters Grade 9 in September as she turns 14. Taken in the context of "squad," she is typical, exhibiting all of the usual highs and lows of life as a young teenager.

But during the Fall as the first semester begins, her parents start to notice things. Marks slipping, irregular sleep, moodiness—outside the norm of what a normal 14-year-old experiences.

They have some theories, but they suspect, at the bottom of it, there is simply too much screen time. They find a way to restrict access to the Internet in their home from 2 am to 6 am. As expected, one morning Olivia says that she hasn't been able to access the Internet. The experiment in limiting access bore fruit: It's now clear that she has been accessing her devices very late into the night.

When confronted, she says, "Well, I need my phone beside my bed as my alarm clock. Sometimes I get texts late at night, and someone is always on Snapchat at that hour. So it wakes me up." Dad pauses and makes a mental note to buy a $10 alarm clock like he used to have.

As the discussion progresses it is obvious that the very last things that her eyes see at night and the very first things her eyes see upon waking are the seemingly infinite social media message from her friends—often with a long list of videos and other content to consume. Sounds familiar to her device-wielding parents, who follow the same pattern.

Mom and Dad think it's time for a family conference. As an open house, they discuss everything. The discussion is the phone: How it is used, the content that it contains and the applications that are used.

What is learned is a shock even for these cool and "with it" parents.

The litany of content that arrives in full living colour into her personal supercomputer is gut wrenching and mind blowing. Even to the most liberal minds, the conversation, content and tone is a quantum leap beyond what was available even five years ago when their eldest daughter was that age.

The easy task (they thought) was the discussion about screen time—specifically, its restriction and appropriate times. That alarm clock excuse is gone and the phone disappears from one hour before bed time until breakfast.

The biggest clue for her parents was that Olivia was chronically tired, more than just the typical stuff that all growing teenagers exhibit.

Sleep

So her parents decided to focus on one problem at a time, figuring the good "choices" that they always told their daughter to make would come from a well-rested and focused mind. They did some homework.

One of the books they found was from Jean M. Twenge. Her book, *iGen: Why Today's Super-Connected Kids Are Growing Up Less Rebellious, More Tolerant, Less Happy—and Completely Unprepared for Adulthood—and What That Means for the Rest of Us*. It was a profound read.

The books summarized a set of compelling and disturbing facts for Olivia's parents about what the impact of lack of sleep was having on Olivia. It is also hinted a suspicious possible cause for it.

- Many teens simply are not getting enough sleep; many are getting two hours less per night than the recommended nine hours in the teenage years.

- 57% more teens were sleep deprived in 2015 than in 1990.

- The increase from 2012 to 2015 was most dramatic: a 22% increase in the number of teens failing to get at least seven hours of sleep.[76]

But here's the real kicker: They learned that the timing of this trend co-related almost perfectly with the arrive of our good friend the smartphone and, more specifically, the arrival of the ubiquitous new social media applications geared to young teenagers like Snapchat.

The chart below says it all.

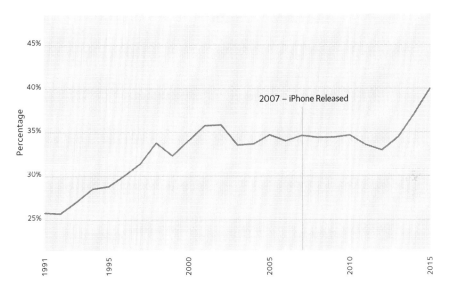

That Crazy Teenage Brain

Olivia's parents knew that this was really important stuff and recognized they were not talking enough about it to other parents or to health professionals. They knew that the teenage brain undergoes dramatic change during this time. All of the big changes that were happening at this moment with Olivia and her friends was connected to brain health in later life. The most important parts of the brain for planning, prioritizing and controlling impulses are the LAST to form and they are greatly impacted by changing brain chemistry and are most certainly affected by sleep changes. The science was clear:

> *All the big changes the brain is experiencing may explain why adolescence is the time when many mental*

disorders—such as schizophrenia, anxiety, depression, bipolar disorder, and eating disorders—emerge.[77]

Forget about content (which we will get to in a second) sleep deprivation seemed to be critical. On further searching, what became apparent to Olivia's parents was that much of the data relating to the impact of smartphones on teenagers was limited and only just beginning to resemble citable and peer-reviewed studies.

Smartphone operating system makers started to get ahead of the impending PR nightmare by putting in "night mode" that reduced the amount blue light emanating from the screen. But deeper research told them that this was not nearly enough.

Focus

The other thing Olivia's parents noticed was the lack of reading capacity, which was likely linked to her difficulty in focusing on one task at a time. She hasn't read more than a book a month for two years. "No one does." she says when queried.

Once again Olivia's parents started searching.

In the excellent essay, "Hamster Wheels—or Why Technology is a Blessing and a Curse" by Achim Walter, the author notes:

> *Creating new things, counteracting negative developments, helping to reduce hunger, disease and suffering: these all require time. Time for difficult development work—and time for reflection… A flash of inspiration is electrifying—but electricity can be dangerous too. A study that hasn't been replicated reliably can give a false*

image of reality. A hastily posted piece of information can trigger a shitstorm.

In our era of constantly available information, we are no longer used to waiting. But waiting, checking and questioning are essential, both for the approval of potentially harmful chemicals and for the development of ideas. This is the only way to put ourselves in another person's shoes, consider alternative plans and rigorously think through positions. Creative freedom is no longer a given; we need to make space for it ourselves. The hamster in us must learn to control its wheel.[78]

Olivia's parents were partly seduced by the call from Olivia and her friends that "everyone does it." She pleaded that she was able to do many things at once and that it was practically "impossible" to do homework with social media and other media present. But her parents knew that there simply wasn't the evidence to prove this—one way or the other. It was a constant discussion with their youngest daughter.

One of the better articles they found was "*4 Things Millennials Need to Navigate the Fourth Industrial Revolution*" by Emerson Csorba. The author looks into the implications of the "hyperconnectedness" of our modern world—especially for young social minds. He clearly evidences that we need space in our lives to create new ideas, thoughts and be creative. Not merely for quality of thought, he says, but to really learn how to create our own perspectives. A key quote from the article that resonated with Olivia's parents—so much that they read it together with Olivia and was food for thought for at least one extended evening dinner-time conversation:

Our hyperconnected world creates challenges around thinking for ourselves, in spaces of temporary isolation from people around us. One of the most important—and

> *challenging—questions to ask in life is "What do I think about this?" ... This social autonomy is a challenge when hyperconnection allows us to connect with both friends and internet users who feel they can provide us with worthwhile advice*[79].

Further on in the article, Csorba hit a more fundamental point from Olivia' parent's perspective and went to the heart of the discussion. In it he discusses that in our hyperconnected world we are busy doing very little. What he calls "superficial engagement". And it's actually exhausting! Combine this with the sleep issue just discussed and we have the recipe for a tiredness that Olivia's parents we starting to even partially appreciate. They started to ask, "Why weren't others talking about this?"

> *Our world of hyperconnection also leads to psychological exhaustion, borne from constant and yet superficial engagement with information. In a world that favours doing, it is commonplace to arrive home from work having done much and yet accomplished very little of substance. This is precisely because work of meaning requires immersion in an activity, often over hours, uninterrupted by others' demands. Superficial engagement, as I find it, leads to an unrewarding tiredness—one of consuming information, which smartphones favour. Knowledge is more difficult to arrive at, precisely because it requires sustained engagement on a topic, with little disruption.*

Larger studies of all ages on the use of technology that our brains—while certainly evolving—are still on the linear track of evolution and physiological advancements. Some extraordinary research is underway in trying to fully understand the brain's plasticity and ability to change at rates faster than it has traditionally. However, the brain's desire for the dopamine hit of the easily consumed "bits" of information that are the hallmark of our wired age is following a reptilian and ancient cry. The mode switching (let alone the far simpler "task" switching) exacts a tremendous price on our ability to critically think and deeply consume concepts that require piecing together the abstract and to pattern recognize.

There are those who say that, in fact, we will use technology to solve this problem. Elon Musk's latest venture, NeuraLink,[80] looks to create machine links into the brain so that we can ultimately keep up with what we are creating. Remember my comments earlier about how we "over-estimate" technology in the near term. The consensus is that these ideas may happen but in the meantime we are experimenting with our kids. We need them to be able to focus on hard problems. And they can't anymore.

For Olivia and her parents, it was the beginning of a journey that they would undertake together during the next 10 years of her life. They were determined to understand and discuss these issues. While they were not alone, they soon discovered that they were in the minority. It gave them pause.

Sleep and focus were symptoms of a bigger unknown problem but the content conversation that followed hit them in the collective gut.

Content: *The Un-Joy of Sex*

Of all of the challenges Olivia faces as she enters Grade 9, her relationship and to understanding of sex is perhaps the most consuming and potentially threatening. Our great social experiment with always-on technology would not be complete without a discussion of pornography and the availability of explicit sexual content.

But it doesn't start or end there. It begins with a culture that has created the impossible choice for young Olivia. In our digital and imaged-based media centered on magazines and online social media accounts of women, Olivia has been given the choice of looking like the pictures she sees EVERY day of her life of hyper-sexualized, photoshopped, perfectly proportioned women, or being invisible and ignored. If there is one thing that has not changed in the history of young teenage girls it's the desire to fit in, to be part of the crowd, to be "lit" in the latest parlance.

That is just the pop culture reality. Then comes the killer: Internet-enabled pornography. As Dr. Gail Dines from *CultureReframed.org* says, the porn industry combining with the Internet has created the perfect storm of "affordability, anonymity and accessibility" (the "three A's of pornography"). The statistics she presents in her brilliant and gut-twisting Tedx talk, "Growing Up in a Pornified Culture" are staggering. Some examples:

- In 2011, it was reported that over two in five (44%) Australian 9–16 year olds had seen sexual images in the past 12 months. This is much greater than the 25-country average of 23%.[81]

- Between 2008 and 2011, exposure to porn among boys under the age of 13 jumped from 14% to 49%. Boys' daily use more than doubled.[82]

- In 2016, a study of 1,565 18–19-year-old Italian students[83], 4 out of 5 stated they consumed pornography. Almost 22 per cent (21.9%) reported that it became habitual, 10% stated that it reduced their sexual interest towards potential real-life partners, and 9.1% reported a kind of addiction.

- In 2017, a Swedish study reported that nearly all respondents (98%) had watched pornography, although to different extents. 11 per cent were found to be frequent users (watched pornography one or more times per day), 69 per cent average users (at least once a month up to several times a week, but less than once per day), and 20 per cent infrequent users (less than once a month)[84]

- In 2006, 35 per cent of Dutch children aged 8 to 12 had had a negative Internet experience in the home, involving an encounter with pornography.[85]

- Well over two-thirds of 15–17-year-old adolescents have seen porn websites when they did not intend to access them, with 45 per cent being "very" or "somewhat" upset by it.[86]

- Fully one-third of all internet traffic—whether measured by downloads, visits or bandwidth utilization are for porn. There is currently more traffic to porn sites then to Netflix, Amazon and Twitter combined!

Whilst the literature varies in its ability to show if pornography directly causes mental health issues, or if, conditions are correlational (existed prior to viewing), or a combination of both, studies indicate that porn users experience:

- higher incidence of depressive symptoms

- lower degrees of social integration

- decreased emotional bonding with caregivers

- increases in conduct problems

- higher levels of delinquent behaviour

But as Olivia enters the second half of Grade 9, her experience with the **Great Social Experiment** hits with full force when she discovers the effect of the "three A's of pornography" with her newest boyfriend. She had had brief "things" with other boys in previous years, but the winter of her Grade 9 year taught her the harsh lessons of pornography's impact on the young man that she has fallen so hard for.

His steady diet of accessible, affordable and anonymous pornography and Olivia's steady diet of pop culture hyper sexuality for girls is a nasty hormone fueled cocktail.

When Olivia's mom found the text messages and Instagram pictures from the increasingly demanding pleas from her boyfriend for semi-nude and nude photos, the floodgates opened.

The parents desire to give their teenager space and freedom was met head on with the **Great Social Experiment** realities. Her parents, accustomed to the experiences of their older daughter had simply not understood the exponential shift in sex norms and had not paid enough attention.

The conversations were challenging at first—tear filled, stubborn, defensive and finally open and truthful. This simply was a world that none of them understood at all.

What was the medium-term impact? What kind of boys would she continue to encounter as she grew up?

The **Great Social Experiment** without baselines or data in its totality would not be revealing its secrets anytime soon.

All of these issues are coming together, however, in the one area that has been consistent throughout the ages: the pressure to fit in as a teenager:

Peer Pressure

A new report, #StatusOfMind, published by the Royal Society for Public Health in the UK[87] examined the effects of social media on young people's mental health. After surveying almost 1,500 people between the ages of 14–24, their findings offered a clear picture of how different social media platforms impact mental health issues, including anxiety, depression, sleep deprivation, and body-image. One tool they used to understand impact on mental health across the various platforms was a survey. Here is the result for Instagram:

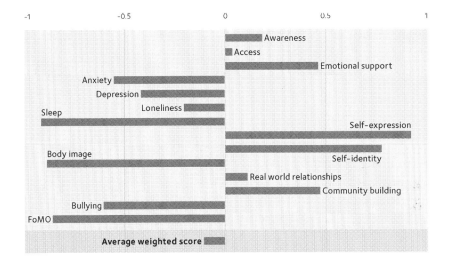

Instagram was found to have the most negative overall effect on young people's mental health. The popular photo sharing app negatively impacts body image and sleep, increases bullying and "FOMO" (fear of missing out), and leads to greater feelings of anxiety, depression, and loneliness.

Olivia was an avid user of both Snapchat and Instagram. One day in the late spring of Grade 9, she came home as usual after school. But she was in a deep funk. After much cajoling (and some of her favourite food), she opened up.

"My picture of the party last night only got 232 likes," she said. Her parents, with a total of 200 friends between them on their social media tool of choice, were confused. "Isn't that a lot?" Dad innocently chimed in. "My best friend got over 400," Olivia wailed.

The parents let the matter drop until the next week when they happened to look at one of the latest pictures of Olivia on Instagram. In days past it would have been part of a spread in a men's magazine. Today, it was part of the portfolio of pictures that Olivia had started and was becoming more and more daring and suggestive.

"But Dad, I got 522 likes for this one," exclaimed Olivia when confronted.

And then they had the conversion about the Internet being written in ink and the lesson started to hit home before too much damage was done.

Peer pressure has been around since there have been teenagers. But this was new. Social, mobile, video captures every moment and "amps" up the need to stand out. Supercomputers in their pocket and parent-hidden apps make this a whole new game.

Once again, **The Great Social Experiment!**

The conversations Olivia had with her parents during her high school years were what got her through. The three of them couldn't help but wonder what other conversations were happening among her friends' families, across their community and around the world as the great experiment played out in tens of millions of homes across the Internet-enabled world.

They made it through because she had the grounding of many things—family, the challenges and grit learned from competitive team sports. and understanding and present parents.

Her conversations with her friends and parents were critical. Her school, struggling with simply managing the changing education demands, was of little help.

In her final semester of Grade 11, she had the opportunity with her school to travel to Malawi as part of a project that involved connecting her class with students from the developing world. It was a trip that completely and fundamentally altered her understanding of the real importance of things like social media and pop culture when contrasted with the staggering opportunities and challenges of the African continent.

The visit would change her life. She didn't know it yet but the seeds of discovery were planted by the time she turned 18 and was ready to head into the next chapter of her life.

chapter 9

OLIVIA

AN EDUCATION SYSTEM DISCONNECT

> We must educate our children in what no one knew yesterday and prepare our schools for what no one knows yet.
>
> *Margaret Mead*

As Olivia enters Grade 12, the school system that has served her adequately so far is showing clear signs of being fully disconnected from the future that is arriving in a hurry.

She thinks about the courses she has taken to date and, while the school board has recently begun to add new courses and has occasionally brought in examples of companies and entrepreneurs to help bring real life into the classrooms, they are nothing like what the future needs. As she continues to think about the critical next couple of years in preparing to enter college or university, she (and her parents) start to see the huge disconnect. Coupled with the increasing cost of post-secondary education are huge concerns that Olivia will enter into an expensive, rote education world unable to prepare her for an exponential future.

Her mother shows Olivia an interesting infographic that portrays the real skills and competencies that the world will require in 2025 and beyond, just when Olivia would likely be finishing a traditional college degree and entering the job market. Taken from the article by the World Economic Forum called, *The Skills Needed in the 21st Century—New Vision for Education*, it recognizes, from the areas of design thinking and creative learning, whole new categories of skills. More importantly it correctly identifies the complex interplay among these skills and calls for the teaching of them (and immediate need to teach the teachers) throughout the curriculum. [88]

21st CENTURY SKILLS		
Foundational Literacies	**Competencies**	**Character Qualities**
How students apply core skills to everyday tasks	How students approach complex challenges	How students approach their changing environment
1. Literacy	7. Critical thinking/ problem-solving	11. Curiosity
2. Numeracy	8. Creativity	12. Initiative
3. Scientific literacy	9. Communication	13. Persistence/grit
4. ICT literacy	10. Collaboration	14. Adaptability
5. Financial literacy		15. Leadership
6. Cutural and civic literacy		16. Social and cultural awareness

Image Credit: The skills needed in the 21st century—New Vision for Education. (n.d.). Retrieved December 17, 2017, from *widgets.weforum.org/nve-2015/chapter1.html*

The addition of competencies and character qualities skills was especially important, Olivia's parents thought. But they had a nagging concern about how quickly the entire education hierarchy would be able to adapt and change in the face of such extraordinary upheaval and, more urgently, what role would they as parents play in it? Surely many of the "character qualities" were those that could be encouraged and advanced in the home?

It felt to them that this study and many others were theoretically sound but the slow, linear advances of the bureaucracy machine of education would either kill it or simply not move fast enough.

In the midst of all was Olivia and her age group. When asked by her parents, she said that she felt like she was the sandwich generation. She was being taught in the "old ways" (as she called them) and that the teachers really didn't know about the real world of technology and stuff. And it wasn't simply about social media and the omni-present mobile applications that were baffling to most of her parents and teachers. Olivia and her friends were aware—even at a relatively young age—of the explosion of new technologies around them. They saw it constantly in their feeds and in the sharing of information on social channels. Though they had more typical things to discuss in their social worlds as teenagers, they knew all of this had to have some impact on what they were learning, how they were being taught and how they were being prepared for the future.

Like many of her friends and teenagers before her, Olivia wasn't sure what she wanted to be when she grew up. She fully understood that many changes would come in her lifetime and so wasn't overly worried. Her parents' insistence on sticking with advanced Math and Science—including some computer programming—seemed like a good idea. But she wasn't sure. On the day after her 18th birthday she was accepted into the university she wanted to go to in the next province over. She was excited to begin the journey and that September she did what so many of her friends did and joined the ranks of the university/college freshman.

The year is 2021. Trump was no longer in power but the fundamental shift in global geopolitics that he had inspired had taken its toll on the world. No one had pressed the red nuclear button and the first Global Symposium on Artificial Intelligence had taken place, much to a collective sigh of relief from many of the world's leaders. Automation was eliminating jobs, 25% of all cars on the road were autonomous, new temperature records were set every month, the seasons were becoming homogenous, the Arctic Ocean was now ice-free, creating a global shipping rush for shortened routes across the top of the world to markets. Experiments in universal basic income were starting to return data on its impact, with mixed but promising results. The iPhone 15 was released.

For Olivia, school was a struggle. She was a smart kid, and her parents had given her the sense to pay attention and challenge things, so she survived her first year—but just barely. She started to feel the urgent pull of some of the bigger issues in the world. Just before the end of the school year Olivia headed to Africa to work on a project in Nigeria with a group affiliated with Médecins Sans Frontières.

She visited Nigeria's capital, the planned city of Abuja. She saw the impact of climate change and the rising sea on the coastal city of Lagos, a city of now 25 million people whose coastal developments were being buried under water. And she, once again, came back from a trip to Africa a changed person. She saw the importance of building a better world.

She came away convinced that the best way to do this was through architecture and urban planning. She knew that the future of cities was going to be invented in Africa.

She applied to architecture college in the UK.

chapter 10

OLIVIA

DISRUPTION AND THE NEW WAY OF WORK

Living in one of the world's great cities filled with interesting architecture and structures was a source of daily inspiration for Olivia. London continued to be an amazing place, and studying there was, by and large a positive journey for Olivia. Spending time in such a multi-cultural centre of the world as London also taught her about how increasingly isolated and mono-cultural her home in Canada was in comparison. Her classmates came from over 25 countries and she estimated that over 45 languages were spoken. All walks of life were on constant display in her urban menagerie. And she loved it.

It was exceptionally hard work and, during her journey, the world continued its evolution at a rapid and unrelenting pace. The technology available to the world in 2023 was staggering. After graduation from her college Olivia managed to get a job at a London-based firm as a junior. Just 22, she was increasingly aware that her skills and education were being eroded. She discovered that the career she was looking for was getting rarer than ever.

In her chosen field, architecture, the typical apprentice

opportunities that led to partnership opportunities were increasingly automated. The drafting work was now fully automated. Robots managed all of the drafting workflow and AI was being used to draft initial creative concepts based on machine learning pattern recognition that matched clients' requirements with existing buildings and structures from all over the world. Machines even created initial mockups that were used to show clients the "art of the possible."

The creative work and design challenges were being shifted up performed by more senior partners. It was extremely difficult to learn the basics because humans no longer performed them. Even more challenging, Olivia's firm was struggling with deep business issues:

- Business model transition: How the firm made money was changing, particularly how much people would pay for the firm's services as well as the method of payment.

- Skill changes: The new technology was complex and clients were increasingly able to do much of the work of design and implementation themselves, because the same tools were made available to them, at increasingly lower costs.

- Quality of employees: Global education systems were not adequately preparing junior staff, especially in the areas essential to collaboration, design thinking, cross-cultural integration and cultural sensitivity.

- Virtual space and creation of new teams: Any large project needed to utilize virtual teams from all over the world. For Olivia and her generation this was great, but her leadership team who had grown up in firms that worked primarily on a face-to-face basis struggled with how to deliver a culture of excellence in distributed team environments.

But one day as she celebrated her first anniversary from the firm, Olivia came across an article written six years earlier, in 2017, which hardened her resolve. The piece lamented that her chosen profession of architect often lacked a global perspective. It was ironic, it said, since architecture and its sister profession, urban planning, had the best possible view of complex adaptive systems like neighborhoods and cities. The problem—especially in the United States—was the incremental point of view.

"There are now 15 metropolitan areas in the world with populations over 20 million. Every one of these cities has stunning business districts, but they also have large portions of their populations housed in poor neighborhoods often constructed informally from found materials", the authors stated.[89]

But the line from the article that stood out for Olivia was by the Canadian designer and architect, Bruce Mau. In Future Practice, he translated the challenge for those who design how we live into the starkest possible terms "We are adding one million people a week to the planet and if the average child is 7 ½ lbs., that's seven and a half million pounds of flesh every week," and "the way we will solve this problem is to design new ways of living to accommodate our scale…and we are a million miles away from that."[90]

Olivia realized that her profession was more than simply about building new spaces; it was about being part of the solution to the problems plaguing the entire planet. In 2023, the world had not blown itself up, though it had come close. The fundamental shift to the Fourth Industrial Age was turning out (mostly) predicted. The "best" of what made the human species, especially our generous nature and quest for knowledge and creativity, had managed to keep its slim advantage over the worst. But what made Olivia absolutely light up as she read this old article was that her work specifically and her profession more broadly embodied all of what had to happen if mankind was going to survive.

Architecture was on the front lines of the battle of exponential change. The skills she had acquired in her schooling from those chaotic days as a young teenager in Canada in Grade 9 through her failed attempt at general university, to the skills and tools she had learned at architecture college had not prepared her for the coming world.

She knew that she could help improve these systems by participating in her old high school's new education crowdsourced innovations platform that had been launched the previous year. The initiative was started in response to the growing cry for ideas on how to make the system work better. The collaborative work systems that were central to how her global firm worked could provide both the tools and the expertise to support the public education system. Her firm's partners were immediate supporters and gave Olivia permission to support and pursue the initiative.

As a young professional, Olivia had to take full and aggressive ownership of her own career and skills development. Her firm had to absorb new technologies that changed the very essence of their business. The leadership had to re-invent themselves as leaders, not only in architecture but also in business. Most importantly, they had to understand their role as a part of a complete holistic system. The collective wisdom approach that drove how the firm created new sustainable new ways to live, work and play were among the most collaborative processes in the world. Every project they did started with a social contract signed by all of the constituents. It was an explicit agreement that let everyone know the expectations of the project collective. Olivia understood that the way her firm worked could be a model for others.

Then one day her firm in London merged with a firm from South Africa, a company on the forefront of sustainable development across sub-Saharan Africa.

At the introductory "meet the new partners" dinner in Central London that night Olivia's managing director pulled her aside and said, "I want you to be a lead on the team that connects these companies together. And there's someone I want you to meet...."

chapter 11

THE LESSONS OF OLIVIA
ACTIONS NOW

Many of you reading this book may know an "Olivia." My bet is that 90% of you are one degree of separation from a young girl exactly like Olivia. Perhaps you *are* Olivia or someone like her.

This simple story of her journey over the next 10 years or so should give you pause. Still, her story is a positive one. Born to parents who didn't separate, who listened and were mostly present, who had enough income to make choices and provide a solid if basic education, Olivia got into a good school, was able to connect to a global movement of change, and through hard work and effort was on her way to succeeding in life.

Olivia was clearly one of the lucky ones. I could have easily written any number of other story arcs that were based on much more dire assumptions. I didn't, for the simple reason that if we can't get our collective acts together around the "happy path" stories, how the heck are we going to build a society that is founded on much less solid ground?

With Olivia's story in mind and with a glance over our

shoulder to the many unfortunate people who will struggle with much less, here are some lessons that I take away:

Pay attention for goodness' sake and speak up!

As we conduct our great social experiment, we are playing with fire with our most impressionable members. Some thoughts:

- Some of you are aware of what's happening with the technology in your kids' pockets, but most of you are not. Learn it. Ask about it. Challenge it.

- Remember Gibson's Laws #2 and #3: The genie won't go back into the bottle (this applies to your kids' phones). Taking the technology away is not the answer.

- Remember you are surrounded by "linear" systems of organization that affect our day-to-day lives and especially our kids' lives. Recognize that these organizations are struggling deeply to cope with change from many sides. Help them. Don't wait to be asked. Respect their challenges and the burden of history, bureaucracy and resource constraints.

- Elect politicians who understand these issues.

Education: *Activism Is Here.*

- School systems across the developed world are struggling. Some will figure things out more quickly than others. The hard reality is that privately funded schools, driven by the demands of already active parents writing cheques, are getting there but the pace is unrelenting. You know in your heart that regardless of whether you can afford selective education, a society needs a functioning, relevant and tax-supported education system. Support it and get involved, directly or indirectly.

- Understand the future skills needed to build our world. Don't wait for someone else to teach you.

- Recognize that your kids are the last generation that will progress on the well-worn path of school, more school, work, retirement. They will need to jump on the life-long learning train and you need to support this now.

- Be a role model. Take online learning to upgrade your own skills. Show others the power of globally connected learning. Learn something new online: a new language or a new skill.

- Talk to your kids.

Businesses: *Lead, Follow or Get Out of the Way.*

- There is going to be serious tension between the future survival of your business, the changing and relevant skills of your workforce, and the increasing threats of automation to that workforce. Your workers, especially young ones, are going to need flexibility and your leaders will need help.

- Understand that you and your business need to conjure up a new concept of the bottom line. It will be impossible in the next 10 years to generate profit without taking into account the full cost of delivering that profit.

- Understand the power of explicitly stating how you wish to work, how your stakeholders need to act and what you see your role in the world to be. As we try to sort out the future of democracy, leadership will come from many places. The organizations in which we work can be and should be a source of that leadership and can easily provide guidelines for our political leaders. There is a huge demand for principled leadership from wherever it occurs. Grassroots movements need to come from both the shop floor as well as the boardroom.

chapter 12

MERVIS
EDUCATION DRIVES EVERYTHING

Let's now venture back to Malawi and view this journey with a completely different set of eyes. Mervis is a typical 13-year-old in the rural part of Malawi near the center of the country. The closest city is Mzuzu (population 80,000). In thinking about her future, Mervis recalled:

> *My dream was to make my mother proud because I'm sure she wanted a bright future for me. My father didn't really understand me and did not want me to go to school. Instead he wanted me to follow my brother into the business and support the family farm.*
>
> *But my dad passed away when I was 12 and his relatives came to take away our property. My mom and brother and I were left stranded; there was nothing to do but just leave everything in the hands of God. My mom could not afford paying school fees so I was sent to a public school. Life has been really hard since then; it was hard to cope with others, but I had no choice but try.*

Mervis lived the challenges of being a young woman in Malawi. She could read English well enough, thanks to her school and her mother's nightly ritual of reading her one chapter from the few books they had in their home. Mervis was hugely curious about the world, and she was determined.

The simple but overwhelming challenge for Mervis was that she was a daughter of farmers. Malawi is a small country of only four cities and just 27 other urban centers, and has a long way to go in terms of urbanization. Farming is the basic way of life. Life is hard and education a luxury; it was especially rare for girls to be educated.

A set of sobering statistics faced Mervis as she looked to the near future. They were not promising:[91]

- More than half of girls in Malawi are married by age 18.

- Only 18 out of 100 girls complete primary school.

- More than a quarter of all girls ages 15 to 19 have already given birth.

- Women with more than a secondary education get married more than seven years later than those with no education.

- Girls age 15 to 19 are more than three times more likely to be HIV positive than their male peers.

- 10% of 15 to 49 year olds in 2014 have HIV/AIDS, according to the World Bank.[92]

- 35% of urban households have electricity, compared with only 4% of rural households.

In the last month of Grade 8, Mervis was visiting her cousin in Mzuzu. While admiring and exploring her cousin's coveted smartphone she happened to read about a summer program hosted by an organization called *GirlUp*, part of the United Nations Foundation. The program was called the **2017 WiSci STEAM Camp,** and it was coming to Malawi. It soon was all she could talk about. She knew this was her chance! When she returned home she told her mother and anyone who would listen about what she had seen.

She describes what happened next:

I had just become the president of my class. So I was at least well known in my community school. I was shy but this little thing of being the class head had helped my confidence a lot.

One day the school announced that a group of volunteers from an education group in the United States were coming by to our school to visit (I didn't know really who they were). I was lucky because as class president I was the one who greeted the group when they came into my class. Three days later, my teacher pulled me aside and said that one of the donors was interested in my talents and wanted to know if I wanted to go to a nearby larger town with a scholarship to learn at the International Academy.

This was the best day of my life and my mother was so very happy. I was sad for a moment when I thought about how my father would have reacted. My mother cried so much because she did not expect a miracle like this to happen

Because of this scholarship invitation, Mervis was also invited to be one of 100 young girls from across Africa and the United States to take part in a summer camp to advance the opportunities for science and math for young African girls.

It changed her life. She read all she could about the program and about STEM opportunities around the world.

STEM AND STEAM

Over the past 15 years, the global community has expended a lot of effort in inspiring and engaging girls and women in Science, Technology, Engineering and Math (STEM) education (and the associated concept called STEAM, which includes Arts as a core element of curriculum). Unfortunately, girls and women continue to face unique and significant barriers in accessing STEM education. According to a study conducted in 14 countries, the probability that female students will graduate with a Bachelor's degree, Master's degree and Doctoral degree in a science-related field is 18%, 8% and 2% respectively, while the probability of male students is 37%, 18% and 6%. While women represent 40% of the global labor force, the staggering evidence and real experience is they are paid lower salaries—sometimes significantly—than men in similar jobs and are concentrated in lower skill, lower wage jobs and industries, with significant gaps in higher value added jobs in STEM fields.

These are the current statistics from the developed world. Mervis soon learned the more sobering data about the impact of education in developing countries like Malawi: [93]

- If all students in low-income countries left school with basic reading skills, 171 million people could be lifted out of poverty—the equivalent of a 12% drop in global poverty.

- If every child in low-income countries completed secondary school by 2030, income per capita would increase by 75% by 2050 and poverty elimination would be brought forward by 10 years.

- An individual's earnings increase by about 10% for each additional year of schooling: rates of return are highest in poorer regions such as sub-Saharan Africa, reflecting the scarcity of skilled workers.

- The return for each dollar invested in education is more than US $5 in additional gross earnings in low-income countries and US $2.50 in lower middle-income countries.

- Girls who quit are three times more likely to be infected with HIV than girls who remain.

- A recent study showed that every additional year of secondary school significantly reduced the chances of individuals later contracting HIV.

Mervis knew she had been given an opportunity that very few of her friends had ever been given. She knew that she needed to do two things: stay in school and find a way to connect into the world of technology, science, the Internet and everything else. She knew she had been given a gift, and because of the way her mother raised her, she felt duty bound and inspired to be not only a role model, but an agent of change as well. She believed with her acceptance into the new school and her attendance at the STEM Camp later that summer, she would begin her improbable journey.

Soon after attending the camp she turned 15. When she

wrote in her diary about her camp experience, it was clear that she knew she could indeed make a difference. She could see that she had a chance to escape the heart-breaking trajectory of her forebearers and many of her friends and peers. This was when she decided that she wanted to grow up to be...

An architect.

Through her studies, the time she spent at the camp and conversations she had with her older brother, she discovered that the future of Africa was to be found in its transition from rural to urban. She learned in her Grade 10 year that the Malawian government was slowly starting to put in place the necessary infrastructure to help in that transition. STEM was her path forward, technology was her vehicle and her world needed new buildings, new cities. Urban planning, design and building would be desperately needed in the coming decades. Mervis had found her path. Meanwhile, a new trend was happening in her country and around Africa that would make this new world possible.

chapter 13

MERVIS

THE NEW TECHNOLOGY REVOLUTION

As Mervis entered her final year in high school, some amazing things were starting to happen in the world around her. Technology, especially mobile, had exploded in her part of the world. The number of mobile broadband connections had reached over half a billion by 2020 in sub-Saharan Africa, more than double the number at the end of 2016, and now accounted for nearly two thirds of total connections in the region. Africa was making the leap into the next generation of Internet-connected devices more quickly than anyone had thought. As predicted by many, the continent was literally by-passing the two decades of Internet and telephone infrastructure build-out that developed nations had had to go through. The pace of innovation was accelerating.

The developing nations of Africa were entering their transition from agricultural rural to modern urban at a time when the all of the plumbing of the Internet was understood, the costs reduced by a factor of 100 since its inception 30 years prior—and the capabilities were staggering.

In 2020, Internet traffic had grown **twelvefold** across Africa as a whole since 2017.

Meanwhile, another equally important development, fueled by the arrival of this new mobile infrastructure, had taken hold.

Banking for the Unbanked

Since 2015, the entire payment system across Africa had changed. Mobile money had advanced from traditional payments such as domestic remittances and airtime top-ups, and could now provide access to more complex financial products, such as savings, credit and insurance. The expanding mobile money ecosystem also offered new opportunities to governments, businesses and individuals.

One of the most pronounced effects of mobile money was that millions of individuals and businesses who had never had access to credit were now able to generate a transaction history, borrow money and pay it back through their mobile phones. Mobile credit was particularly prevalent in sub-Saharan Africa, driven by the ease and accessibility of the loans, high demand from the middle-income population, its instant nature, and a relatively mature mobile money industry.

This fortuitous combination of access to technology and financial freedom came at a crucial stage in Mervis' life. She had the presence of mind (thanks, mom!) to appreciate that, while she didn't have all of the answers and was still a poor farm girl from rural Malawi, this was a once-in-a-generation opportunity and she was determined to take full advantage of it.

African Realities

But Africa is not the developed world. Innovation and socio-economic transformation was met head on by some staggering facts regarding population, poverty and demographics:[94]

- Of the world's 7.3 billion people, 1.2 billion are Africans, representing the second largest in population after China. By 2050 the population will have grown to 2.2 billion people. Of these, two thirds are considered youth under the age of 25 years.

- The median age in North America is 37 to 40 years, while in Europe it is 37 years and projected to be 52 years in 2050. Africa's median age is 19 years.

To compete, what Africa requires is an empowered youth population that is clear on its self-identity, self-worth, and possesses an inspired commitment to create and change. If managed correctly, this is the power that can potentially put Africa at the centre of global innovation.

Hopeful stuff. But there is an often missed reality. Mervis came across a powerful essay by Awel Uwihanganye, the Founder and Senior Director at the LéO Africa Institute, who dug deeply into the issue. His essay reminded Mervis that we can't simply bring in the latest technology and drive a new economic revolution. Why? Because there are many regions in Africa where the unemployment rates—especially among youth—are staggeringly high; regions where poverty and simple survival are paramount. He argues for a balance of the best of the new economy being met with the building of new efficient industries where labour remains a factor of production.

But it was his final point that really hit home with Mervis:

One can argue that the age of disruptive innovations is another way for capitalism to keep a stranglehold on world economies, and keep maximum profits flowing in the pockets of the already rich. The mindset towards African innovation must offer solutions to the most pressing problems or needs of the people in critical areas such as health, education, finance or even recreation. Efforts at such innovations must also be invested in, and be facilitated to reach scale. Investing in innovations that are not inclusive, that exclusively cater for the elite and well to do, will only heighten the income divides with negative consequences.[95]

The changes that needed to come to her country had to come from those who lived within and understood the challenges. Young people who were simply running away to find a better life or those importing made-in-the-rich-world solutions that simply tilled the soil of inequality were not what the country needed. Mervis knew she was part of the new solution and Awel had inspired her.

There was other evidence of a disconnect from the promises and the realities of modern Africa. Despite the remarkable growth in technology and innovation across the African continent, many Malawians remained offline as Mervis began university. A disproportionate share of these unconnected individuals came from underserved population groups, including women and those on low incomes, who still face significant barriers to mobile Internet adoption. According to the ITU, women in Africa were 23% less likely than men to have access to the Internet in 2020. This gender gap in Internet access is underpinned by a persistent disparity in mobile phone access for men and women. In short, there was still a long way to go.

But there was a final piece in the puzzle that would push Mervis into this new world.

Role Models, Community and The New Architecture

The research indicates that few women get exposure to entrepreneurial role models in their formative years, with only 15% saying that they definitely had family and friends who often talked about business when they were young, and only 29% that there definitely was a successful business owner in the family and extended family.

The lack of a role model carries through into women's careers and later lives, too, with only 14% of women reporting that they have a business mentor or role model.

"Young women need to be exposed to the possibilities and the benefits of having their own business at home, in their communities and schools, and in the media," says Joanne van der Walt, the Sage Foundation program manager for Africa.

Mervis knew a few people that had owned small businesses but she was never able to easily seek out their advice or mentorship. She really didn't know anyone in her community who could teach her about the larger world of business.

She read as much as she could. But things changed when Mervis was accepted into the Malawi Polytechnic School for architecture in Blantyre, the country's financial and commercial hub.

In the course of her studies, she was invited to visit a project in Chimpamba called the Legson Kayira Community Center & Primary School. This extraordinary project resulted from a collaboration with Youth of Malawi, a New York City-based charitable organization, and the South African firm Architects for a Change (A4AC). Built in 2016 and expanded over the past five years, it had expanded from a small schoolhouse to a community centre that was loved by all.

There, Mervis learned of the new collaboration that was at the forefront of social change in Africa. Backed by North American expertise and capital, designed and manufactured in South Africa, and built and assembled on site with Malawian labour, the school changed the community forever. It would also serve as the final inspiration for Mervis on her journey.

On that day, she met one of the founders of A4AC, Deirdre, an architect, who was working on the roof of the new addition at the school. Covered in dirt and sweat, Deirdre laboured under the hot Malawian sun assembling the new solar panels and Internet drone uplink. She looked down and smiled at the Malawian girl with the serious face and keen smile.

chapter 14

MERVIS

BUILDING THE NEW AFRICA

Mervis was enthralled by this partner from the South African architecture firm. When Deirdre came down, she sat with Mervis and they talked about all of the things that were happening in the local school, in the rest of Africa and around the world. The architect told Mervis that, when she was ready, she should reach out. The school that Mervis was at had just connected up to the new B4A ("Broadband for Africa") node that was providing 200mb access to the global Internet, and her computer was one that had been donated by the world's largest architect firm in partnership with Google, which had recently acquired the world's leading design firm, Adobe.

"Send me an email when you get back," offered Deirdre. Mervis didn't have to be asked twice. They corresponded throughout her senior year. Deirdre was able to offer advice and became a formal mentor to her. She was part of a global program developed by MIT called the Venture Mentoring Services for Africa (VMSA) that connected female professionals with girls in university from the developing world to help navigate the critical last mile in their journey from school to work.

As she was preparing her application for an intern role at Deidre's firm, Mervis came across the inspiring essay, "7 Architects Designing a Diverse Future in Africa," by Dario Goodwin. Written in 2015, it had predicted the world of 2025 in Africa and the much of it had come true. Goodwin's belief in the importance of a native African industry of urban design and architecture that respected the uniqueness of the land and its people and rejected the naïve colonial view inspired Mervis as she applied for the new role. Goodwin's essay was a turning point for development agencies, not-for-profits and the growing entrepreneurial and innovation class that emerged in Africa after 2020.

His rallying cry gave birth to the "#RightWay" movement that took the best of the emerging sustainable energy and development practices hardened in the late 2010's in the developed world that were imported with cultural sensitivity into Africa's urban development boom.

> Many say… that the future indisputably lies in Africa. Long featuring in the Western consciousness only as a land of unending suffering, it is now a place of rapidly falling poverty, increasing investment, and young populations. It seems only fair that Africa's rich cultures and growing population (predicted to reach 1.4 billion by 2025) finally take the stage, but it's crucially important that Africa's future development is done right. Subject to colonialism for centuries, development in the past was characterized by systems that were designed for the benefit of the colonists. Even recently, resource and energy heavy concrete buildings, clothes donations that damage native textile industries, and reforestation programs that plant water hungry and overly flammable

trees have all been seen, leaving NGOs open to accusations of well-meaning ignorance.

Fortunately, a growth in native practices and a more sensible, sensitive approach from foreign organizations has led to the rise of architectural groups creating buildings which learn from and improve Africa. Combining local solutions with the most appropriate Western ideas, for the first time these new developments break down the perception of monolithic Africa and have begun engaging with individual cultures; using elements of non-local architecture when they improve a development rather than creating a pastiche of an imagined pan-African culture.[96]

In the fall of 2023, Mervis became the new intern for A4AC in Johannesburg. Her journey had taken her from the farm to the city and had beaten the odds. And she knew it and never forgot.

Throughout the next years, Mervis moved ahead, though not without continued struggles. As she became a full-fledged architect, she still felt the challenges of being a woman in a field dominated by men. She worked with people all over the world, yet still found it difficult to change her self-image and that of others that she was the minority, from a "provincial" backward part of the firm.

Then something amazing happened: Deirdre became the managing director of the entire operation, which had grown to over 200 people across 10 cities in Africa and had a new office opening in London in 2024.

The firm had recently partnered with a new London firm on a major new project in the Malawian capital city, Lilongwe. The project was such a success that the firms decided to merge operations.

The newly combined firm had a new company mission: Reimagining Africa! It became a beacon for new opportunities in the firm and Mervis had been asked to join the team.

But all wasn't perfect. In 2025, Africa's urban renewal was only just coming off the drawing board. In some countries, strong, democratic governments provided the catalysts for investment in critical infrastructure and new cities, backed by investments by healthy combination of global and local private capital, international agencies and local public sector support. These successes relied on a complex interplay of many variables. But most important were the presence of strong local governments, their commitments to new forms of education and private capital's investment in the long-term growth of the local innovation ecosystems.

A4AC was a leader in its vocal support of this partnership model for its projects. As the firm's CEO, Deirdre vociferously rejected the idea of her firm being involved in projects that didn't connect all of the pieces of the complex urban re-redesign so needed in large African cities. In her speech to the Architecture for Africa conference in London in 2025 she rallied her fellow urban planners and architects to the cause. She specifically called out the projects that invested billions in new planned cities that ignored the great challenges of building the long term social fabric of Africa. She called for a new social contract for her fellow professionals that explicitly called for a new partnership with strong, democratically elected governments and patient capital and rejected the simple, short term view.

She implored her audience,

> *The future of Africa is in the way we create our cities. We have a chance to bring the best of the world's technology, urban planning and design with the unique magic of the African way of life. We have a blank canvas. What picture are we going to paint?*

Over the past few decades, Africa indeed had begun to move away from the tired stereotype of the shift from poor rural to a magical urban transformation. Incredible investments had been made across the continent and the was much success in connecting the best of the African cultures with new urban designs. But it was a generational challenge and many were not up for the long term task.

Deirdre called out the past two decades of experimental "new cities" with gleaming towers, smart city infrastructures and massive shopping malls built from scratch. She noted,

> *While these trophy projects appear beautiful at first, they are not our future. They bring a false vision—a mirage—to the vast majority of our people who will only set foot in them as low paid wage slaves to those who can afford it. They can't live here! They can't play here! We know from the history of the western cities that urban centres that provide benefits to all were the ones that survived. Let's be better than this!*

Deirdre spoke to the crowd and concluded, "We must change the way we think and work. Technology, collaboration, science and education must be fused with best practices, sustainability and respect for cultures. Our firm is going to change the way we think of the world around us. We will build a new social contract with all with whom we work."

Her speech was the catalyst for a fundamental change with her firm, her team and for her profession.

Sitting at the dinner table that evening in the heart of London, Mervis was overwhelmed with excitement and emotion—and a tinge of fear. Deirdre had wanted her to join the London office and be the liaison with the African team. Her mandate would be to help teach her new UK partners the best of what the African continent could be; to help deliver on the firm's mandate to change the continent one project, one city, one region and one person at a time.

She then introduced her to the woman who was going to lead this from the London side. An ex-pat from Canada.

Her name was Olivia.

chapter 15

THE LESSONS OF MERVIS
THE NEW AFRICAN EXPERIMENT

Many of you reading this book don't know a Mervis. My bet is that 90% of you are more than two degrees of separation or more from a young girl like Mervis.

Her story, like Olivia's, is simple, and I could have chosen any number of other story arcs that were based on much more dire assumptions. But I chose her storyline for the simple reason that there *will* be leaders from Africa who will be positioned to change the world. When they emerge we must encourage and support them, and help their stories come true.

So, with Mervis' story in mind and with a glance over our shoulder at the many unfortunate people who struggle with much less, here are some lessons to take away.

What Happens in Africa is the Future of Our Planet.

Since for most of us much of the continent of Africa is many thousands of miles away, literally and figuratively, it is hard to fully appreciate its importance. It is my hypothesis that in the next decade or so, throughout the many different countries and regions of this continent, we will be watching our species' proverbial canary in the coal mine. I have a warning similar to the one I gave Olivia's parents in the previous section: **Pay attention.**

As noted at the beginning of this chapter, Africa is not a single entity. The Western world (developed countries and its citizens) is guilty of oversimplifying this extraordinary place. I focused on one girl in one small province in a very poor part of Malawi to tell my story. As with Olivia's journey, I could have chosen any number of paths. Her journey is a narrative that can give rise to thoughtful action.

We are mostly certain that the branch of Homo sapiens that evolved to our current seven billion started in Africa. Over millions of years, our earth has changed much, but it is now headed for a showdown measured in decades. In the ultimate existential irony, I believe that our survival and future—be it in two decades or 100 millennia—will be determined in the very same place.

Mervis' story is a journey that we can imagine, and we can and must influence its outcome. We can provide the opportunity for her success. Bluntly, her survival is the challenge of the human species. We must:

- Help create sustainable living for a new billion people

- Make access to education available to all

- Make technology accessible and affordable

- Provide women equal rights

- Rid the governance structures of corruption

These are the challenges of the upcoming decades. My story of Mervis is one of optimism. Her journey will be a key part of the solution if it can be repeated 10 million times, creating leaders and role models within a new global social contract of hope.

A message to these two amazing young women:

I can't wait to meet you, Mervis. I am glad that you met Olivia. The two of you represent the future of our species. (No pressure.) But you are not alone.

Our journey together into the near, real future begins with something called a New Social Contract. In the final part of the book we will give your parents and future colleagues a kick start (or perhaps simply a kick!).

My ask of them will be to create this new agreement—perhaps many of them. And while it will be started by your parents and their contemporaries, the real social contract will be written by you two remarkable women and those whom you will influence and call friends, and those who advance because of the work you do and the role models you have become.

It will, I hope, be the rallying cry and manifesto that will call upon us all to look hard at the problems of today and say, "Not on my watch."

PART 4

TOWARDS A NEW SOCIAL CONTRACT

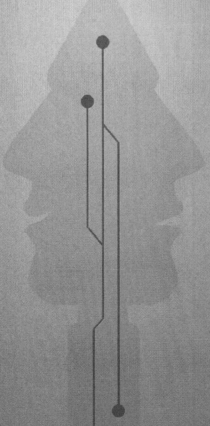

chapter 16

A NEW DEFINITION OF INNOVATION

We have now looked at the Tip of the Spear from both ends of this fast moving and species-defining javelin. We have felt its impact by imagining a journey over the next 13 years of two remarkable young women about to run with it.

The tip of the spear that is poking at our sides is being driven by realities that I have coined The Three Laws of Disruption. They have created an untenable situation for our planet and now the Tip is glowing white-hot as it speeds into the unknown.

The tail is the crucible of inequality that has been around since the dawn of humanity and is gathering force at a speed and intensity never before seen.

But before we look to solutions, we need to look closely at a couple more things.

Are things really that bad?

Many authors argue that we are just about to enter an era where exponential technologies decades in the making are about to provide us with the one thing we all crave: abundance.

Abundance: The Future Is Better Than You Think, by Peter Diamandis, speaks of such a world. It is a fascinating read. It goes to the heart of mankind's assertions from eternity that we must conserve and plan for scarcity. The hardwiring in our brain struggles to believe that we actually have all that we need—and more—to build the world of our dreams. The cost of the energy needed to power our lifestyles and to feed ourselves, he argues, is following those curves we discussed at the beginning of the book. They are exponentially bending towards zero.

It is a fundamentally foreign though, isn't it?

What Diamandis argues is that this future will give us more of the one thing that is truly fixed and the one thing we have constantly run out from the dawn of time. The gift of time itself.

> *Each of us starts with the same twenty-four hours in the day. How we utilize those hours determines the quality of our lives. We go to extraordinary lengths to manage our time, to save time, to make time. In the past, just meeting our basic needs filled most of our hours. In the present, for a huge chunk of the world, not much has changed. A rural peasant woman in modern Malawi spends 35 percent of her time farming food, 33 percent cooking and cleaning, 17 percent fetching clean drinking water, and 5 percent collecting firewood. This leaves only 10 percent of her day for anything else, including finding the gainful employment needed to pull her off this treadmill. Because of all of this, {Matt} Ridley feels that the best definition of prosperity is simply "saved time." "Forget dollars, cowrie shells, or gold," he says. "The true measure of something's worth is the hours it takes to acquire it."*[97]

The staggering examples in Diamandis' book and others of how much better off we are reminds us of just how far we have come. I cannot disagree. The "abundance" that Diamandis and others describe is, to any student or even a casual observer of the world around us, just over the proverbial horizon.

I certainly believe in the power of exponential change curves. I also believe, however, that classic economic theory based on scarcity cannot predict the future. The model is broken or soon will be. On a positive note, our discussion of how we overestimate technology in the near term and underestimate it in the long run comes significantly into play here. I think it is fair to say we simply don't have a clue about what it means to plan our actions today based on a future we can't understand. As someone once said, forget about the known(s); we don't know even know what we don't know!

So while Diamandis and others believe in a future of abundance, I am more cautious. My concern is that we might not get to reap the benefits that are in front of us if we don't first confront the fundamental journey of our hearts. And it requires a profoundly different definition of what it means to innovate as a species.

THE INNOVATION OF WAYS AND THE INNOVATION OF THINGS

The purpose of the last part of this book is to look hard at these questions and to present ideas for what I think the human species needs to be doing to keep the metaphoric javelin flying straight. There is genuine hope for our future tempered with real concerns about the flight of the spear and where the tip will land.

I have spent much of the past two years meeting and

talking with people about their thoughts, fears, ideas and reactions to the technological changes happening around us. Between slugs of beer, coffee and other substances, I have searched for clues and patterns of hope and solutions. The conversations have been superb.

These tête-à-têtes have been essential to help me understand whether or not this stuff really matters—and to whom. What I can conclude, however, is that the best way to augment this exponential growth of technology is through connected, "present" humans in ways that inspire. Joy, compassion, love, hope and kindness are extraordinary human traits. When magnified by a movement, they are truly inspiring, and indeed exponential, easily matching the pace of technology.

In a provocative and inspiring TED talk called "The Dream We Haven't Dared to Dream,"[98] , Dan Pollata notes that, while human ingenuity has exponentially increased the transistors on a chip over the past 40 years, we have not applied the same exponential thinking to our dreams or human compassion. As he says, "We continue to make a perverse trade-off between our future dreams and our present state of evolution."

Some describe this ethical stasis as "the tyranny of the OR" (versus the power of AND). I call it the battle of the Innovation of Things versus the Innovation of Ways. Can we create a society that values the advancement of *how* we work, live, play and learn as much as it values the advancement of technologies that we can buy to make life easier?

The Innovation of Things is the creation of new tools, toys and devices that serve to make human life better, easier or more interesting. It is the cars that move us, the computers that connect us and so much more. It is the "What."

The Innovation of Ways is the use of these tools to change the way we do something, from the way we build our cities, to how we share rides or rooms to how we run our finances or build and lead our organizations. It is the "How."

In the history of innovation, How most often follows What.

The tension between the power of technology to create new and magical things (the What) and the ability of the human species to understand and absorb it all (the How) is central to this book's thesis. It is a tension growing more acute each day.

Much of the innovation at the white-hot edge of technology disruption is centered on the Innovation of Things: Robotics, AI, VR/AR and the long list of exponential technologies noted at the beginning of this book.

In my opinion, we are suffering because the Innovation of Ways is lagging—in many cases, significantly.

In my research it has helped me to break down innovation into Innovation for **wants** and Innovation for **needs**.

There are different dynamics of capital flow, timeline horizons and even the types of people who are involved in these two different areas.

Using my "Tip of the Spear" analogy, it kind of looks like this:

	Innovation of Things	Innovation of Ways
Wants	TIP (Exponential)	
Needs		TAIL (Linear)

In a market- and capital-driven system, the return of capital in the top left quadrant likely has the shortest path to success and highest return in the medium-long-term. Anecdotally, it has often been said that the earliest use of the newest technologies is often the adult porn industry, i.e., new "toys" being applied to the most obvious and basest of wants. Just look at what's happening in Virtual Reality and think back to the introduction of VHS, the Internet and many others throughout history. Gaming is a close second.

But this is far too simplistic in this extraordinary time of innovation and technological advance. We are seeing new devices that are the result of combinations of exponential technologies in all areas of our lives. Many innovations start in the upper quadrant, then move down into the bottom left. Facebook starts as a commercial tool and ends up providing a simple, cheap, and user-driven method for not-for-profits to create and feed a community of interest. More recently (and still in planning stages), autonomous vehicles start as a way to create competitive differentiation in the car business but have begun to move into proving self-driving services to those who are unable to drive—the blind, infirmed and others.

What about the Innovation of Ways? As Dan Pollata might say, we lack the same heat and light in the areas of changing how government, the education systems and heath care work. These are clearly the "ways" of life—how we govern, organize, teach and heal.

The optimist in me says that the winners in the Innovation of Things will begin to turn their attention to the Innovation of Ways. That is why, for example, I am so interested in what Mark Zuckerberg is doing with his *Internet.org* initiative to connect the entire planet online.

A cynic would say that he is doing this simply to add more users to Facebook. But on very close observation and analysis—and stopping short of speaking directly with Zuckerberg himself—that is clearly not the case.

A pessimist would say that we have stalled in our view of global thinking. Nationalistic tendencies will stop us from transferring the Innovation of Things to the Innovation of Ways as we build walls literally and figuratively across the world. The new titans of technology will continue to earn their deep and concerning concentrations of wealth and power.

This struggle and debate is the essence of the *Tip of the Spear*, and within this discussion lie some of our paths forward. If the energy that powers our lives and feed us is becoming free, I am certain that our new energy imperative is to take some of the riches, abundance and time that have come from these extraordinary human advances and translate them into reimaging the Ways of life.

The optimist in me emerges and it is time to look more closely at one such initiative that has given me a clue to where we go from here.

chapter 17

THE RAINFOREST AND THE HUMAN COLLECTIVE

> Compassion is the radicalism of our time,
> *The Dalai Lama*

> Innovation comes not from the basic ingredients of economic production, but from the way that people interrelate to combine and share ideas, talent, and capital. A community that facilitates such relationships is a biological system we call a Rainforest. Its animating process is creative reassembly.[99]
> *Victor Hwang & Greg Horowitt.*
> The Rainforest: The Secret to Building the Next Silicon Valley

As we look deep into the challenges that result from the inequalities of technology, wealth distribution and opportunity, we can be easily overwhelmed. These are really big problems, made much worse by the growing number being brought into the online, connected universe for the first time.

Two thoughts occur to me.

The first comes from Yuval Noah Harari's book, *Home Deus: A Brief History of Tomorrow*. As Harari looks into the fundamental questions of both the current day and humanity's future, he digs deeply into the very essence of what makes the human species so successful:

> *Over those 20,000 years human-kind moved from hunting mammoth with stone-tipped spears to exploring the solar system with spaceships not thanks to the evolution of more dexterous hands or bigger brains ... Instead, the crucial factor in our conquest of the world was our ability to connect many humans to one another. Humans nowadays completely dominate the planet not because the individual human is far smarter or more nimble-fingered than the individual chimp or wolf but because Homo sapiens is the only species on Earth capable of co-operating flexibly in large numbers. ....*
>
> *History provides ample evidence of the crucial importance of large scale co-operation.*[100]

While the honey bees and massive multimillion-member ant colonies all appear to co-operate, Harari argues that they do so "inflexibly," according to some ancient hard-wired evolutionary constructs.

Humans are unique because of their ability to connect *flexibly*. Our wiring of the planet is, according to this logic, predictable. If our human evolutionary path led us to dominate the planet, the Internet was entirely obvious. While arguably for exploitation and power, our exploration of the planet was really just about communication and collaboration.

The first question, then, is "Will our collective infrastructure simply evolve or does it need guidance and direction?"

If the answer is that we ought to allow the human species to evolve naturally, then I worry that we are on a trajectory—a "Tip of the Spear" issue—that will see our advancing technologies and inequalities causing further global divides. In other words, we might blow ourselves up before we can evolve enough.

More on this later.

The second question is at the opposite end of the communication scale: One to one.

"Now what!?"

The Rainforest

I live in an amazing part of the world. Some may think of Canada, with its reputation for tolerance and activism in diversity and global consciousness, as the "new hope" for our troubled world. If you can stand the winters, it is truly a wonderful place to live. In the province of Alberta where I live, however, we have been struggling to diversify our resource-based economy. Alberta is the oil and gas powerhouse of Canada, a global leader in conventional and oil sands production of hydrocarbons. Alberta produces a significant portion of Canada's GDP and, through our nation's system of equalization payment, benefits "have-not" provinces.

In other parts of the world (several Nordic countries, for example) where energy revenues made up a disproportionate part of the economy, countries were able to establish rainy day funds to compensate for the gradual decline of oil and gas revenues. Alberta has not done this very well. We have spent today's fortune without putting anything away for the future. Making things worse, there are many highly educated climate change deniers in Alberta. This is an educated place. But, as the old saying goes, "If all you have is a hammer, everything looks like a nail."

What I generally do when having such discussions is to say "believe what you will, but there is no denying that there is a "time box" on hydrocarbons for our ever-growing energy needs." This could be in 20, 50 or even 100 years. But if our experience of exponential change and technology discussed in Part 1 is any hint, **change at this pace is slow at first and then is staggeringly quick.**

My guess is precipitous change in my lifetime. I am 57.

So, what does this have to do with the Tip of the Spear and one-on-one change? The answer is found in the way we have been introducing change into Alberta's stubborn information technology ecosystem and economic diversity conversations.

There is an urban legend about how this all started: Sometime in 2014 or 2015, Justin Reimer, one of the new Assistant Deputy Ministers in the Alberta government, read a fascinating book: *The Rainforest: The Secret to Building the Next Silicon Valley*. Written by Victor Hwang, currently a thought leader at the Kaufman Institute, and Greg Horowitt, the book outlines a comprehensive, multi-disciplinary approach to understanding how complex economic systems are connected and how they emerge as successful innovation hubs. Most importantly the book provides a framework that others can use to think about and advance the maturing of their ecosystem.

The deputy minister responsible for Alberta's innovation policy was enthralled. Word had it that anyone who came across Minister Reimer would be met with "Have you read this book?"

One of the people who happened across the minister was Brad Zumwalt. Now, in the scheme of global successful technology entrepreneurs, it is unlikely that Brad's name would be mentioned. But in Alberta, Brad is one of the most completely successful entrepreneurs we have ever produced.

Zumwalt's multiple exploits in the digital image industry were exceptionally profitable and successful. Yet it was what he did next that has defined his remarkable record. Instead of buying toys and travelling first class around the world, Zumwalt used his newfound wealth to create Social Venture Partners, an organization that brought venture investing to not-for-profits. It was capitalism meeting the real world of struggle and inequality.

In the language of *The Rainforest*, Brad was an ecosystem "keystone." He took Reimer's advice and read the book.

And anyone who came across Brad after he read the book would be met with, "Have you read this book?" I was one of those people, and I read that book. Cover to cover.

Timing is everything. As Alberta saw its fortunes begin to change for the worse, the technology sector began to stir. An economic pipsqueak beside the huge energy industry, the tech sector here lives in a very different world. As an energy province, Alberta is global player—if not in tier one, definitely in tier two. As a technology sector, however, we are fifth tier. This is not a criticism, simply a fact. As such, we do our best to move the needle. We are not ones to waste a good commodity bust and provincial-wide recession to help our capitalist system move forward into the digital economy.

Back to Brad, the book and big change.

Brad and I decided to do what good entrepreneurs do: start something quickly, refrain from looking for handouts, and get others onboard to support our efforts. We convened a gathering of the best new economy thinkers, entrepreneurs and government leaders and asked the authors of the book to walk us through their *Rainforest* vision. It has been a remarkable journey and continues as of this writing.

Our biggest take-away? Something the authors call the **Social Contract**.

The Social Contract

Hwang and Horowitt spend a significant amount of time on the concept of culture. Primary research told them that individuals in an innovation ecosystem act in ways that traditional economists looking at complex economic systems fail to adequately pick up.

Actors in an ecosystem behave atypically. Most economic

models posit that individuals behave rationally to maximize *individual return on investment,* whereas the successful Innovation Rainforest is populated with people whose activities and behaviours maximize *return on involvement.* A new type of ROI:

> *Almost anyone in a Rainforest who has started a company has witnessed this new ROI firsthand. It is the successful entrepreneur who provides you with an hour of free advice. It is the advisor who drafts up a three-page strategic plan for you, completely on his own initiative and with no expectations. It is the engineer who starts working on your software project without knowing what her actual pay is going to be.*[101]

This new behaviour is a critical aspect of what makes innovation ecosystems thrive and is a huge clue to how we need to reshape our interpersonal behaviour in this new world.

Again from the book:

> *Whereas social contracts in our ordinary lives are implied and invisible, we seek to make Social Contracts for the Rainforest manifest and visible. These contracts are physical documents that attempt to manifest the Rules of the Rainforest, as applied to local circumstances. They are like written constitutions for innovation ecosystems: they provide maps for people to navigate the informal rules of their new communities. As in cognitive behavioral therapy, Social Contracts work by making people more self-conscious of their choices for action.*[102]

In Alberta, we created our own version of the Social Contract, which reads as follows:

SOCIAL CONTRACT

RAINFOREST

I am joining the Alberta community of people sharing a common faith in continuously growing the CULTURE of innovation and entrepreneurship. We are an inclusive, silo-busting, sector agnostic, all industry, open source, ego shrinking, ecosystem building, entrepreneur focused, wide open, social barrier smashing community.

This agreement describes what it means to innovate together. It defines what we allow and pursue versus what we resist and avoid. It is a litmus test for the quality of our decisions, actions and ways of being together. By signing this agreement, I agree to join the Rainforest community, to embrace and promote this explicit set of 10 values and principles:

1. **DIVERSITY:** I embrace diversity, strive to create equal opportunity for every person and I am open to meeting anyone in this community.
2. **FREE HELP:** I understand that I will receive valuable help from others for free.
3. **TRUST:** I will give trust to others before expecting to receive trust in return.
4. **PAY IT FORWARD:** I agree to "pay forward" whatever positive benefits I receive.
   - For every introduction I get, I will provide an introduction to another person.
   - For every hour of advice I receive, I will give an hour of advice to someone else.
   - For every risk someone takes with me, I will take a risk with a different person.
5. **FAIRNESS:** I will be nice and treat everyone fairly. I will take advantage of no one.
6. **LISTENING:** I will bring people together and listen, as none of us is as smart as all of us. I believe in the big tent. I believe we gain strength through diversity.
7. **HONESTY:** I will be truthful and frank. I will break rules and call out elephants in the room.
8. **TEAM SPORT:** I will create teams to play, dream, experiment, iterate, and persist.
9. **I UNDERSTAND MISTAKES HAPPEN.** Failing quickly and cheaply are acceptable ways of testing new ideas.
10. **SHARING:** I will open myself to learning from others. I am eager to act to learn. I will share my knowledge in the spirit of the Creative Commons to help nurture learning in others.
11. **ROLE MODEL:** I will lead at times and follow at other times. Each person acts as a role model for everyone else. I will live these ideals and enforce them as a member of the Rainforest community.

SIGNATURE _____ DATE _____

If you read this social contract carefully you will see an explicit set of behaviours that is categorically opposed to the zero-sum game of hardcore, winner – take-all, market-only capitalism.

Most importantly, it makes explicit a set of commitments to which signatories are bound, creating a constitution of sorts that sets forth how everyone in our ecosystem is to behave.

Back to Harari: If we are collectively successful as a species because we have found the recipe for flexible collaboration, my experience in Alberta has given me a clue to the new operating guide for this connected planet. Clearly, we need to create a culture of innovation if we are going to survive, but it must begin with a culture of trust.

If I imagined a new Social Contract as the guiding force by which a new, globally connected, social media system would abide, what would that look like?

Imagine if we could wipe the slate clean. What if our exponential technologies could help us implement a global social contract that made explicit what we are trying to accomplish?

Further, what if we began by simply building agreements between two or more people on how to behave? Simple human behaviours, coded and amplified.

How would this new social contract read? Try this:

- Converse online as though your reader were looking you directly in the eye.

- Use language and words that originate from respect.

- Say two positives before you say one negative.

- If you don't understand the issue, don't comment. Two ears, one mouth.

- Ruthlessly create, curate, and promote sources of truth.

- Learn to identify what truth is.

- Truth is truth; facts are facts; opinions are just that. But if they are not respectfully articulated, call that out.

- Well-reasoned, fact-based, alternative world views are a new gold standard and the true currency of trust.

I know many of you are asking how we can engineer this change globally. Aren't there simply too many negative people in the world who would overwhelm our attempt?

Perhaps, but social movements are like the exponential curves we discussed at the beginning of the book. We see slow-changing and minute movements at the beginning, but all of a sudden they can explode. I have seen this happen in real time. I have watched a group of 10 or 20 dedicated and committed people grow into 100s and now 1,000s. Without pushing an agenda. Without advertising. Simply with sharing solid messages and the good work of people trying to make our ecosystem better.

I can also imagine that in the near future the social technology that has now enveloped the planet will be part of the solution; it will help the conversation and advance a social contract.

Perhaps, for example, instead of signing up new friends on Facebook, Snapchat or Twitter, the acceptance of a friend was in fact the acceptance of an explicit social contract between two people or a group of people. Instead of some complex and opaque "terms of use agreement" with Facebook, they could create a simple, human, understandable social contract and ask two "friends" if they wish to adhere to it. Creating a new friend on Facebook, a new follower on Twitter or a new link on LinkedIn starts to grow up and starts a to fell a little more like the deeper meaning of friends or a business colleague in the "real" world.

Perhaps we need to challenge our technology platform companies into leading this charge. As I'm writing this, the US Senate is bringing the leaders of Facebook and Twitter to account for their role in the suspected Russian involvement in fake advertising and news in the 2016 presidential election. What new social contract might we suggest these CEOs bring forward as a new operating guide for a connected planet? More affirmatively, should we demand it?

Perhaps someone is already putting together a new platform based on an explicit approach to dealing with awful human behaviour and the subversion of the most powerful conversation tool we have ever invented.

Perhaps we should simply stop using the tool until we figure it out? Ah, but that genie never quite fits back in the bottle, does it?

The idea of a new social contract sprang from the idea that a culture of trust can (must, in fact) be built if we are to create the conditions for change and innovation. A culture of trust quickly eliminates barriers to disagreement, i.e., the post-fact world. The beginning of a new turn recognizes that the unique success of humans is based on our ability to effectively collaborate *at scale*; we now have the tools to make this global.

We just got off on the wrong foot. After all, the Internet is really only 20 years old.

A new social contract is the future of innovation. I learned this directly with the amazing Rainforest volunteer group here in Alberta that was inspired by the groundbreaking work by Hwang and Horowitz. I am convinced its time has come.

Let's dig deeper, make it personal and apply this vision of

explicit social contracts to the many roles we play in our lives. And let's see if we can look at our own individual behaviours as a counterweight to the laws of disruption and increasing inequality in the world. That world can be the relationships we hold dear, the places we work, the countries in which we will and the planet that inhabit. Big or small, the sum of our relationships can change the planet.

chapter 18

SOCIAL CONTRACTS FOR THE REST OF US

> To discover the rules of society that are best suited to nations, there would need to exist a superior intelligence, who could understand the passions of men without feeling any of them, who had no affinity with our nature but knew it to the full, whose happiness was independent of ours, but who would nevertheless make our happiness his concern, who would be content to wait in the fullness of time for a distant glory, and to labour in one age to enjoy the fruits in another Gods would be needed to give men laws.
>
> Jean-Jacques Rousseau, The Social Contract

The social contract is a powerful historical construct. The writings of Jean-Jacques Rousseau are required reading for those trying to understand the power of these explicit contracts of fundamental principles.

As we begin identifying solutions to the Tip of the Spear problems, it is clear that some constituents need to re-think their implicit and explicit social contracts with people who matter within that specific context. My experience in the work we have been doing with the Rainforest initiative in my home province has profoundly influenced my thinking. As of this writing, over 900 leading participants in the digital innovation economy have signed the social contract document discussed in the last chapter. It has been brought inside the largest public and private organizations. It has defined working relationships in nascent start-ups. It has changed the conversation, both in tone and in content, amongst the young and old.

For example, we have met every Wednesday for the past 14 months—attracting up to 90 people a week—with each week seeing many newcomers arrive. We give people a chance to introduce themselves and we ask them to answer three questions: "What do you do? What do you need from the ecosystem? What can you give to make the innovation system better?"

The permission to ask for help within a social contract that asks one to give back, is a powerful message, isn't it? It gives a new company or individual the confidence to be very clear about what they need and to not be shy about asking for it. But this is done in a spirit of fundamental trust that the community will also ask, "What can you give back?"

Most importantly, it has shown me a real-world example of the power that explicit commitments can have on the players who make complex change happen. In short, the social contract is a keystone tool that could be used across many Tip of the Spear issues.

The examples that follow are a best-efforts summary of the issues and challenges facing a broad array of constituents that have surfaced in the Tip of the Spear discussion. It is by no means a complete list, nor does it represent the fullest treatment of the issues. But my hope is that I may give you an idea of where to start if you choose to dig deeper.

Time to get involved!

PARENTS

Parents are the frontline of interaction with those who will be most affected by the coming changes. If my tale of Olivia and Mervis resonated, you will see that the journeys ahead are perilous for those coming of age in the coming decade. In these times of change, it will not be enough to simply count on goodwill, good policy or good luck to see your way through. As a parent you will have to take assertive control over the ways in which technology impacts your children.

Parents are in an extremely difficult position. For many, the very Tip of the Spear threatens their way of life. Their jobs and future advancement are in doubt as automation continues its relentless advance into an increasingly broad and deep set of skill and work. Simply understanding the technology their children use and the content they consume is a staggering challenge, as my chapter on Olivia reminds us. It is a tough time to be a parent.

Some advice:

Learn for yourself and understand that this discussion matters. Read about the future and the trends. Discuss the issues with your kids. The endnotes of this book provide a suggested reading list for parents. Check *www.thespear.co.* for my top 10 reading list and for examples of how other families have opened and framed the discussion of technology and the issues raised in the book.

If you have younger kids, re-read my story of Olivia and have a conversation with your children about sex and communication and focus and the challenges of the coming world. Become a digitally smart parent by creating a new social contract with your children.

Become digitally literate. Stop looking at funny cat videos on Facebook or reading the 12 ways that avocado can help your sex life on Twitter. Then have your 13-year-old explain Snapchat to you.

Far more importantly, talk to your kids about the skills their future will require and what they can do today to prepare for tomorrow. (This, of course, will require you to understand what those skills might be.)

Demand that your school bring in realistic role models from

the *near* future. For example, ask your kids' principal about what they are doing to showcase entrepreneurship as a career.

Make a point to have similar discussions within your school parents group. Better, create an online collaboration space. I have created one on my site that you can join, or you can create one for free with one of the widely available collaboration tools—Slack is the one I use (*www.slackhq.com/*). Use it to share articles and books, and to start discussions around the challenges. For example, I have created five topic areas ("channels") in my Slack space:

- **Technology**—What's Next (links to great, factual and easy-to-understand articles about future technology)

- **Kids**—Discussion on specific challenges with kids of all ages

- **Skills & Education**—Discussion on 21st century skills

- **Resources**—People or organizations involved in next-generation skills development, development sciences, etc.

- **General Discussions**

Ask your child's teachers what they understand about the future of Artificial Intelligence and its impact on learning. Ask if they understand that an always-on digital mobile device is hindering your child's ability to think original thoughts, be focused and put together a cogent sentence.

Forward them the excellent essay on the subject entitled: "4 things Millennials need to navigate the Fourth Industrial Revolution"[103] to get the conversation flowing.

On page 175 is an example of a social contract that might work for a typical family.

SOCIAL CONTRACT

MY FAMILY

This agreement describes what we mean to live together as a family. It defines what we allow and pursue versus what we resist and avoid. It is a litmus test for the quality of our decisions, actions and ways of being together. By signing this agreement, I agree to make this family a better unit by embracing and promoting this explicit set of values and principles:

1. **DISCOURSE:** As a family we will talk, frequently, with an open heart and open mind. Sometimes we will talk when we don't want to, but we will follow this Social Contract when doing so.
2. **MANY GUIDELINES; FEW RULES:** Few things in life are absolute. We can guide each other without making us all the same. The rules are here in this document; guidelines will come and go and are negotiable. The rules are not.
3. **TRUST:** I will give trust to others before expecting to receive trust in return.
4. **CONSEQUENCES:** Do what you say. Always. If you don't, understand there will be consequences for your actions, both for you and for others.
5. **FAIRNESS:** I will be nice and treat everyone fairly. I will take advantage of no one.
6. **LISTENING:** I will bring people together and listen, as none of us is as smart as all of us. I believe in the big tent. I believe we gain strength through diversity.
7. **HONESTY:** I will be truthful and frank. I will break rules and call out elephants in the room.
8. **BEING DIGITAL:** I will make a consistent effort to understand the tools we all use every day. Not just in what they do, but the need for limits and the need to put down the tools at least once a week for a few hours.
9. **SHARING:** I will open myself to learning from others. I am eager to act to learn. I will share my knowledge in the spirit of the Creative Commons to help nurture learning in others.
10. **ROLE MODEL:** I will lead at times and follow at other times. Each person acts as a role model for everyone else. I will live these ideals and enforce them as a member of this family.

SIGNATURE _____ DATE _____

VOTERS

One of the most common "ah-ha" moments during discussions I have had in researching this book is the realization that our 200-plus year experiment with democracy is on the verge of collapse as it deals with the onslaught of change and disruption. I don't need to spend much time documenting this. You see it every day as you catch up on your news from media sources that are becoming increasingly unreliable information partners and with the nationalistic, frightened populism spreading into many corners of the world.

As noted previously, it is ironic that just as we wire up our planet with the perfect vehicle for a global conversation and massive collaborative decision making, we are letting the quality of our discourse, so vital for a healthy and functioning democracy, deteriorate.

As a voter—especially in the United States where the electorate is subject to brutal multi-year campaigning—it has become extremely difficult to sort through the chaff of political discussion.

Anecdotal evidence is backed up by many thoughtful articles on the future of democracy, its current crisis and solutions to keep the process functioning. From the paper, "E-democracy in Action" by three partner organizations—Cooperation assembly (Estonia), Open Knowledge Finland (Finland), Public Participation Foundation (Latvia)—the authors document and lament the decline.

> The evidence for these claims {of democracies decline} stems from many observations that we are constantly reminded of by the headlines—party dealignment; low voting turnout on any elections; lack of trust in politicians; harsh disparities between different social groups, and rising economic inequalities. When it comes to the reasons behind this, opinions will vary—you will hear

people say that the very nature of democracy is flawed and that a certain change is in order. Some will blame the liberal-capitalistic state arrangement where too much attention is paid to the individual and wealth accumulation (with which, arguably, comes the power and thus more influence among the general public, which is something you don't really want in a democracy).

But no matter what the true reasons behind this disengagement and passivity, we are left with a general public which seems to be more apathetic than active and more dismissive than critically engaged.[104]

So what to do as a voter? Well, for me, and for the authors of the case studies found in the paper above, it is clear that citizen engagement, especially in between traditional electoral cycles, is critical. The paper outlines nine different case studies of varying degree of success where the process of engagement was enhanced through various technologies such as crowdsourcing. There has been much promise, to be sure.

Having worked directly in the digital conversation and crowdsourcing space, I can tell you that it certainly is a remarkable tool for connecting thousands of people from across a city or across the world. It is tempting to imagine voter engagement tools based on next-generation crowdsourcing tools to rapidly connect the electorate on short-, medium- and long-term decision making.

The reality is that "warm" human conversations (that is, those resulting from heated political, social and economic debates found in typical day-to-day civic conversations) require special handling in the "cold" technical medium of online conversation tools. We humans are quick to opine. But when we lack the more complete social cues that are found in face-to-face conversations, we require moderation and digital referees. My experiences have told me we forget these base human responses at our peril.

But as the authors summarize,

All of these are all examples of democratic revival on a digital terrain. And sure, most of these initiatives are still operating on {an} ad hoc basis, but they nevertheless represent the wider urge to participate somehow and prove that the general public is by no means incapable, disillusioned or passive. Quite the contrary—they cannot wait for you to give them another megaphone to raise their voice on.[105]

As a voter, there has never been an easier or more important time to get involved. Be part of the solution.

So what to do?

- Ask anyone running for office or any party to explain their position on the future of innovation.

- Be wary and vigilant about fake news and the spreading of untruths. Curate and cultivate your own sources of truth and widely share them with colleagues.

- Send your candidates copies of the Fourth Industrial Revolution paper from the WEF (or give them this book) and ask if it resonates with them and why or why not.

- Or get more specific about a disruptive issue relevant to your jurisdiction—for example, job re-allocation in the world of AI; new training and skills development for displaced workers; universal high availability Internet; tax policies to attract the necessary capital to make a lot of this happen.

- Or (and this is a technique I have used) ask them how they feel about having a Minister of the Future and what that position would do.

SOCIAL CONTRACT

VOTER

This agreement describes what we mean to be an active and informed citizen in the world of democracy. It defines what we allow and pursue versus what we resist and avoid. It is a litmus test for the quality of our decisions, actions and ways of being together. By signing this agreement, I agree to live these as part of my ongoing commitment to participatory democracy

1. **DISCOURSE:** I will listen to other viewpoints as often as I can. I will not always agree with other points of view but I do control how I react to them.

2. **CIVILITY:** I will always speak to those on opposite sides of an issue as I would have them speak to me or other members of my family. When faced with incivility, I will disengage: after sending a copy of this Social Contract.

3. **EDUCATION:** I will take time to learn about the issues locally, regionally and nationally. I will not comment on issues online or in person until I am aware of the issues.

4. **VOTING:** I will vote in all elections. I will be committed to ensuring all members of my family do the same.

5. **FAIRNESS:** I will be nice and treat everyone fairly. I will take advantage of no one.

6. **LISTENING:** I will bring people together and listen to discussions about issues of the day. None of us is as smart as all of us. I believe in the big tent. I believe we gain strength through diversity.

7. **HONESTY:** I will be truthful and frank in my discussions.

8. **PARTICIPATION:** I will actively participate in the electoral process. This will range from supporting the electoral process as a volunteer to actively running for office or supporting someone who is.

9. **SHARING:** I will open myself to learning from others. I am eager to act to learn. I will share my knowledge in the spirit of the Creative Commons to help nurture learning in others.

10. **ROLE MODEL:** I will lead at times and follow at other times. Each person acts as a role model for everyone else. I will live these ideals and enforce them as a member of my community.

SIGNATURE _____ DATE _____

On page 179 are some elements of a simple voter's Social Contract that we should require everyone to think about and, ideally, read prior to being able to vote:

CIVIC LEADERS

I have the utmost respect for people who put up their hand and say, "I am going to run for office." I don't care if it is for local school trustee or for a national political role. There is great sympathy and empathy for those who deliver on the commitment that comes with public service. I remain sympathetic to and optimistic about these people (at least upon introduction) and I think the vast majority of them enter into politics with good and sincere intentions. However, the system pollutes and corrupts many. I am optimistic, not naïve. For those who do put up their hand, I have some thoughts.

While much of politics is local, I really believe that the next 10 years requires a new form of leadership. As I often say: "Managers *manage within* a paradigm; leaders lead *between* paradigms."

For many years, politics has been about managing within known and existing paradigms. As citizens and voters we have been asking our politicians and elected officials to be good "stewards" of the economy and democratic process. I believe, however, that those days are rapidly coming to a close.

No matter what level of political office you are seeking, there are significant leadership and paradigm-busting issues ahead. As a new or incumbent leader, think about how you can change the tone and content of the conversation. In this era of great change and great opportunity, it is wise to be aware of and sympathetic to the impacts of these changes and to reflect them directly into the promises and vision you are crafting.

Understand the meaning of these new social contracts and ways of living. Some suggestions:

- Be transparent. Learn and use readily available digital tools with your constituents. Don't simply farm out the "digital stuff" to a millennial (or younger) volunteer. Live digitally and live it authentically.

- Learn and adopt best practices from around the world.

- Understand how to harness the power of the crowd, digitally, without inciting the power of the mob.

- Create hackathons that invite smart people to team up with average citizens and civic leaders to ask how to make government better.

- Speak authentically about the realities and challenges of this "new order."

While there are many, many issues that need to be addressed in our local, provincial and global contexts, the issue of inequality is the one that crosses all of these boundaries. So, for the civic leader, a discussion on the new economic principles that will need to be driven from leadership.

Specifically, we need to have a discussion about Universal Basic Income.

ECONOMIC DISRUPTION AND THE NEW INEQUALITY:
UNDERSTANDING UNIVERSAL BASIC INCOME (UBI)

Universal Basic Income is an economic theory and practice that has as its central thesis that the overall social health of a population, and thus society as a whole, increases when each individual in that society is provided—without qualification—with enough income from the state to provide for basic needs.

If you think political debates are tricky, wade into a conversation about UBI with a politically conservative colleague and watch the sparks fly! Trust me, I have and it is fascinating stuff and not for the faint of heart.

As we enter into Tip of the Spear territory, many leading writers such as Thomas Freidman (*The World is Flat* and *Thank you for Being Late*) and Tony Atkinson (*Inequality*) and technology thought leaders like Space X/Tesla's Elon Musk, Apple's Tim Cook or Amazon's Jeff Bezos, are looking straight at the challenge of wide-scale disruption of the job market. They have paid at least public lip service to the crushing reality that mass automation will play in the job market of the near future and that something must be done. All argue that it will *not* come from the existing classic economic thinking of market rule and supply and demand.

It is clear for me that, whether it is universal basic income (UBI) or something similar, we must address the fundamental change in what constitutes work, jobs, income and wealth distribution in an era of exponential change in technology capabilities, global inequality and human population. Solutions emerging from an informed, global debate and discussion in the next 15 years will be a species-defining process.

After many debates, I, for one, will not accept the notion that the "free market" will decide the outcome. The incredible opportunities that the future of human invention will bring will be crushed by the pace and scope of the changes coming to our way of life.

For new and incumbent civic leaders, the expression "all politics is local" is one to keep in mind as you try to take what appears to be a global and complex problem down to the local level.

The debate is urgent.

In *Voting for Freedom: The 2016 Swiss Referendum on Basic Income: A Milestone in the Advancement of Democracy*, the authors describe the preconditions that have led to the call for UBI:

> If we have the capacity to feed more people than currently inhabit the earth and innumerable people still suffer hunger, it is because we treat wealth as if it were rare, a scarce good, a limited resource. What we are missing is the ability to deal in an appropriate manner with the abundance that actually exists. The discipline demanded of us by abundance is generosity. Those incapable of it become voracious or greedy—two modes of behavior in the face of lack. Basic income takes seriously the fact that lack has ceased to exist and frees us from having to feel it subjectively in the wrong place. Whoever sees themselves permanently threatened by material dearth races throughout the world like an animal in search of food. They either snatch their prey away from their own kind, or eat willingly out of their master's hand.
>
> Unconditional basic income grants everyone what they need and invites everyone to show what they are capable of. There is no shortage of products today. What we are short of if we are not to perish in the midst of our abundance is courage and imagination. These cannot be coerced; they can only be enabled....[106]

This is a powerful call to the Innovation of Ways we just discussed. They sound very much like the words from our friend Peter Diamandis in his work *Abundance*. In an online discussion forum, one commenter noted:

> Society will not give up technological and social innovation, nor should it. However, industrially developed nations have failed so far at distributing fairly—that

is, to everyone's benefit—the advances made through rationalization. The result is left-over activities, de-qualification and unemployment for some, and densely compacted work, self-exploitation and exhaustion for others. [...] Basic income will not make anyone lazy, any more than gainful employment makes anyone intrinsically hardworking. Humans beings are, quite simply, beings of activity.[107]

The concept of UBI is a complex one. In the 2016 Swiss vote, UBI was rejected by a 76.9% majority. Germany rejected a similar petition in 2013 on the grounds that it would cause a significant decrease in the motivation to work among citizens, with unpredictable consequences for the national economy.

In a report, the "No" side concluded:

- UBI would require a complete restructuring of the taxation, social insurance and pension systems, which would cost a significant amount of money.

- The current system of social help in Germany is regarded as more effective because it's more personalized: the amount of help provided is not fixed and depends on the financial situation of the person; for some socially vulnerable groups the basic income could be insufficient.

- It would cause a vast increase in immigration.

- It would cause a rise in the shadow economy.

- The corresponding rise in taxes would cause more inequality: higher taxes would translate into higher prices for everyday products, harming the finances of poor people.

- No viable way to finance basic income in Germany was found.

In a contrasting opinion, one author reminds us that we have had many such experiments in the past. UBI is not a new concept. We used to individually draw water from a stream or the community well. We then invented plumbing and faucets. With these new inventions, it was loudly argued that we would no longer be able to learn of the ways of the community and the gossip of the neighbours because there was no more gathering at the water place. The argument was who was going to make sure everyone turns off the taps and they are not improperly used?

> *Unconditional basic income is no revolution...Once basic income has been implemented, it will be just as much a matter of fact as water faucets are today. Unconditional basic income is not an added frill; it is letting go of superfluous stipulations. It lets possibilities run free. People unable to control themselves tend to want to control others. Basic income promotes self-control.*[108]

As an incumbent or aspiring civic leader, you may not believe in UBI. For many, it goes to heart of the "survival of the fittest" philosophy that fuels the harder edges of capitalism.

To these traditional "right wing" politicians or candidates, my call to you is to recognize and believe that the scope and speed of change (remember the three laws of disruption) will be such that market forces alone will not suffice. Disruption will be too pervasive, too sudden, and will affect many of your typical base supporters. The state *must* help shape the response to these challenges; new solutions, not old dogma, will be required. What is your new social contract with those being left behind by exponential change?

To my traditional "left wing" politicians or candidates, the linear forces we discussed at the outset will ensure that governments alone cannot move quickly enough. Moreover, the new paradigm-busting leaders are emerging from the very industries and communities that are driving change. (I have more to say to my digital leaders in the next section!) You need to encourage the market forces to make consumer and ethical decisions for the good.

In summary, I believe there are truly bi-partisan and civic leadership components to a new political leadership social contract that we should require all candidates to sign before committing to run for civic office. The political landscape is very complex today. It is a tinderbox of emotions and tribalism. I dare not trivialize the extent to which it is broken and needs repair and healing—certainly in the U.S. but also around the world. There has never been a more important time (save for the leadership at the dawn of our global wars in the past century) for a new, calmer, reasoned and diverse set of leader voices. It starts, I believe, with a new social contract for our leaders to share amongst themselves that says, "how we behave as human beings toward our constituents, our opponents and our country, starts with us."

DIGITAL LEADERS

Later in this chapter I have included a more general discussion on the new leadership social contract for the broad category of business leaders. Here, however, I want to pay specific attention to the leaders of the digital economy. In this

SOCIAL CONTRACT

CIVIC LEADER

This agreement describes what we mean to be an active and informed citizen in the world of democracy. It defines what we allow and pursue versus what we resist and avoid. It is a litmus test for the quality of our decisions, actions and ways of being together. By signing this agreement, I agree to live these as part of my ongoing commitment to participatory democracy

1. **STRAIGHT TALKING:** I will answer the question asked. I will never be afraid to say, "I don't know." I will speak clearly and will learn other languages of my constituents to be able to share with them directly.
2. **CIVILITY:** I will always speak to those on opposite sides of an issue as I would have them speak to me or my colleagues. When faced with incivility, I will disengage and not feed it: after sending a copy of this Social Contract.
3. **EDUCATION:** I will take time to learn about the issues locally, regionally and nationally. I will not comment on issues online or in person until I am aware of the issues.
4. **HUMILITY:** I will be transparent and honest. By acting in a spirit of humility I will never assume that I know more than someone else. I will also recognize the real challenges and pain that people feel. I serve them in all that is good and not good.
5. **RESEARCH:** I will dig into the issues that matter to those I serve. I won't give trivial answers to hard questions.
6. **LISTENING:** I will bring people together and listen to discussions about issues of the day. None of us is as smart as all of us. I believe in the big tent. I believe we gain strength through diversity.
7. **HONESTY:** I will be truthful and frank in my discussions.
8. **MEASUREMENT AND PERFORMANCE:** I will hold myself and my political establishment accountable to measurable progress and I will communicate this progress regularly.
9. **SHARING:** I will open myself to learning from others. I am eager to act to learn. I will share my knowledge in the spirit of the Creative Commons to help nurture learning in others.
10. **ROLE MODEL:** I will lead at times and follow at other times. Each person acts as a role model for everyone else. I will live these ideals and enforce them as a member of my community.

SIGNATURE _____ DATE _____

group I include the venture capital organizations funding the new economy, the leaders of transformational new companies and the leaders of traditional companies undergoing transformation.

The change you are bringing—through sweat, failure, capital, risk, luck, avarice, greed and all of the other sins and virtues—is not neutral. It has winners and losers. Increasingly though, the fruits of our digital capitalists' labours is greatly increasing the inequality gap. Whether directly or indirectly, it has resulted in people being left behind with a growing despair that they will never be able to participate in the bright and shiny future you are creating. They are growing increasingly cynical of the closed cabal of funding and wealth whose spoils are increasingly going to fewer and fewer individuals and organizations. Worse, these are your customers. Simply stated, will they be able to buy the products you are selling?

This is not Adam Smith's "invisible hand" at work. This has the potential to derail the train of change. Your social contract with the world is profound and important. Your authentic, active and clear voice is needed.

Your most revered colleagues are beginning to understand this.

In a powerful article[109] entitled "How Elon Musk Proved That Thought Leadership Is The New Patriotism," *Forbes* magazine fully grasps the importance of the new captains who have emerged from the leading technology companies. They look back at a famous Battle of Thermopylae (where the expression, "This is Sparta" comes from) for the lesson of how the few can conquer the many. This famous battle, as you may recall, saw a group of about 300 Spartan and other soldiers face down nearly a million Persians (at least according to ancient sources) at Thermopylae. It inspired a thousand years of songs and legends. The lessons for today, concludes the *Forbes* article, are clear:

No, neither you nor I are Elon Musk, and hopefully neither of us will have to lead a spear charge any time soon, but as modern business leaders with insider knowledge and powerful viewpoints, we can impact the lives of more individuals than Leonidas the Warrior King did that day. But we have to speak up.

Today, it isn't Persians who are blocking us at the pass—it's fake news. It's alternative facts. Forces like these, hell-bent on manipulating the truth, have coalesced to make vocal leadership even more meaningful and necessary than it was a few months ago. So you're not a politician? Doesn't matter. Fake news is everywhere, not just in politics; it's infiltrating healthcare, tech, business, and economics. If you have a meaningful platform and you're not sharing your views, you're doing the world a disservice. Elon Musk is already out there tweeting what he believes in. Ask yourself: why aren't you using the platform you have to truly lead?

The leaders of the new economy are, for the most part, beginning to recognize the incredibly important platform that their economic stature brings. As our government leaders struggle with their own shifting social contract issues, into this void will ride the (hopefully) principled leaders of the new economy. Or they will stumble and trip. The rise and fall of Uber's CEO Travis Kalnick is a case in point. Forgetting some of the basic rules of decency and leadership, Mr. Kalnick almost singlehandedly wiped out billions in market value for the ride-sharing startup, launched numerous shareholder lawsuits and got himself fired.

Mark Zuckerberg, the 32-year-old Facebook founder, seems to understand his role and, like Musk and others, has truly begun the transformation from a transactional CEO to global visionary. Setting aside all cynicism as well as

Facebook's late-2017 Russia problems, Zuckerberg has correctly understood his stature and the reach of his platform. At two billion-plus members, the Facebook community is larger than any nation on earth.

Let that sink in for a minute.

In February 2017, Zuckerberg penned a 5,000-word manifesto that took full advantage of his perch. While others will debate the content for some time, the fact that its reach was truly global was not lost on many. To help solve the world's hardest problems, Facebook, according to Mr. Zuckerberg, can contribute by answering five fundamental questions:[110]

> How do we help people build supportive communities that strengthen traditional institutions in a world where membership in these institutions is declining?
>
> How do we help people build a safe community that prevents harm, helps during crises and rebuilds afterwards in a world where anyone across the world can affect us?
>
> How do we help people build an informed community that exposes us to new ideas and builds common understanding in a world where every person has a voice?
>
> How do we help people build a civically-engaged community in a world where participation in voting sometimes includes less than half our population?
>
> How do we help people build an inclusive community that reflects our collective values and common humanity from local to global levels, spanning cultures, nations and regions in a world with few examples of global communities?

These are profound questions, regardless of how you think or feel about Mr. Zuckerberg. As a participant in the digital economy and traditional business creation for 30 years, I fully appreciate the type of persona attracted to the adrenaline-filled world of high-stakes technology. The winner-take-all aspect of that era will likely be over soon. And once again, Gibson's Three Laws of Disruption are simply too powerful to ignore. Those affected by technological disruption are not going to stand for this much longer without leadership. And that leadership won't come from our traditional politics. It will come from those to whom we continue to give our money, privacy, and attention: the technology leaders.

So to those of you leading the charge in building the new economy, here are some blunt words:

- Stop intellectually masturbating about the utopian future and recognize that people are really struggling with this. It is complex and it is frightening to some. Be compassionate and authentic.

- Understand that the rising tide raises all boats. Invest in your communities to support new ventures and entrepreneurs.

- Write about the coming changes with compassion and humanity. Study and think about inequality and the issues of the global community.

- Give back to the community. Help young entrepreneurs understand that technology is always two-sided.

- Spend time with your elderly colleagues. Help them bring out the best of their wisdom in new digital contexts and provide ways for them to have meaning in this new paradigm.[III]

- Speak at local schools and give back. K–12 education is where we will help shape our future economy participants and leaders.

- If your organization is global, think about Mervis and her world. Resist and fight the corruption that plagues the fragile democracies of the developing world. Use your economic weight for real change, both on the ground with teens like Mervis and in the corridors of power.

- Speak out about making our developed democracies smarter and exponential in their thinking and transactional efficiencies. Ask, as President Obama did, why it is easier to order a pizza online than it is to vote online.

- If leadership on the issues of technology disruption and inequality isn't forthcoming, stand up and fill the leadership void and speak truth.

- Do not tolerate untruths and "fake news."

In summary, as Klaus Schwab states in his work, "The Fourth Industrial Revolution," how we react and adapt to technological disruption is in our power to manage:

> We can only meaningfully address these challenges if we mobilize the collective wisdom of our minds, hearts and souls. To do so, I believe we must adapt, shape and harness the potential of disruption by nurturing and applying four different types of intelligence:
> - *contextual (the mind)*—how we understand and apply our knowledge
> - *emotional (the heart)*—how we process and integrate our thoughts and feelings and relate to ourselves and to one another
> - *inspired (the soul)*—how we use a sense of individual and shared purpose, trust, and other virtues to effect change and act towards the common good

- *physical (the body)*—*how we cultivate and maintain our personal health and well-being and that of those around us to be in a position to apply the energy required for both individual and systems transformation*[112]

That's heady stuff. But if the organizations you head are collectively worth over a trillion dollars, they have plenty of capital and time to think it through. If you are one of the digital leaders, you and your competition all need to be leaders in this journey. On page 194 is an example of a social contract I imagine for the leaders of the technology community.

PROFESSIONALS

Perhaps no one in the future will be impacted as much as the professional service provider, who all have a bullseye on them. The laws of disruption discussed in this book are coming for the exclusivity and scarcity that allows value to be attached to knowledge. Of course, the transition to open availability of knowledge and vastly increased access to expertise won't happen overnight. For some professions it has already happened (ask the tax preparer). For others, it is imminent (auto mechanics in a world of simplified electric vehicles). For still others (like the primary care physician), it is many years away.

Regardless of where or when you believe the change will fall fastest or hardest, that change is coming: From the seminal book, *The Future of the Professions*, the authors look into the world of professors, doctors, lawyers and architects (among others) and see that disruption has already begun:

> There is growing evidence that a transformation is already under way. More people signed up for Harvard's

SOCIAL CONTRACT

DIGITAL LEADER

This agreement describes what we mean to be an active and informed citizen in the world of democracy. It defines what we allow and pursue versus what we resist and avoid. It is a litmus test for the quality of our decisions, actions and ways of being together. By signing this agreement, I agree to live these principles as part of my ongoing commitment to leading technology change.

1. **LAWS OF DISRUPTION:** As technology advances at an increasing pace, I recognize that technologies once invented never go back into the labs and organizations and people are struggling to absorb the change my product may represent. I will always answer the question, "Just because we can, should we?"

2. **ACCESS:** Advances in technology create an opportunity to balance the inequality across the country and around the world. I will always ask myself and my team if there is a way to help others with what we are building.

3. **PAY IT FORWARD:** I agree to "pay forward" whatever positive benefits I receive.
   - For every introduction I get, I will provide an introduction to another person.
   - For every hour of advice I receive, I will give an hour of advice to someone else.
   - For every risk someone takes with me, I will take a risk with a different person.

4. **DISCOURSE:** I will take time to engage in the global debate on technology. I will understand that as a leader of change I can be a positive and active voice in the debate.

5. **HUMILITY:** I will be transparent and honest. By acting in a spirit of humility I will never assume that I know more than someone else. I will also recognize the real challenges and pain that people feel. I serve them in all that is good and not good.

6. **RESEARCH:** I will dig into the issues that matter to those I serve. I won't give trivial answers to hard questions. My technology is part of a greater ecosystem.

7. **LISTENING:** I will bring people together and listen to discussions about issues of the day. None of us is as smart as all of us. I believe in the big tent. I believe we gain strength through diversity.

8. **SHARING:** I will open myself to learning from others. I am eager to act to learn. I will share my knowledge in the spirit of the Creative Commons to help nurture learning in others.

9. **ROLE MODEL:** I will lead at times and follow at other times. Each person acts as a role model for everyone else. I will live these ideals and enforce them as a member of my community.

SIGNATURE _____ DATE _____

online courses in a single year for example than have attended the actual university in its 377 years of existence. In the same spirit there are greater number of visits each month to the WebMD network than to all of the doctors working in the United States...meanwhile the Pope has 19.3 million followers on Twitter and the Dalai Lama 10.4 million...[113]

If you are a professional, I am sure you have been feeling the change for a while—though, like the proverbial frog in hot water, I am not sure if you recognize yet that the water is about to boil.

The authors paint two different futures for the professions. The first is where the powers of exponential technologies will enable a vast streamlining of the efficiency—speed and scope—of service delivery. Better, faster, cheaper services delivered in new ways.

The second future is more profound and to the authors more likely. In that future, the professional morphs into a curator and catalyst of machine intelligence and provides new and infinitely flexible solutions to a broader, more diverse and significantly more complex set of problems to new real and virtual clients.

But the issue for professionals is more than simply the preservation of business models or how their transition occurs. It also includes an inevitable discussion about how they manage the league of un- or under-employed professionals, many of whom are in the highest economic brackets of our society and have made significant financial, time and family sacrifices to achieve a level of authority and expertise. This will surely be hard to manage.

But there are profound moral issues that transcend the employment issues. What, for example, are the moral ramifications of using technology to make life-or-death decisions

SOCIAL CONTRACT
PROFESSIONAL

This agreement describes what we mean to be an active and informed citizen in the world of change and innovation. It defines what we allow and pursue versus what we resist and avoid. It is a litmus test for the quality of our decisions, actions and ways of being together. By signing this agreement, I agree to live these principles as part of my ongoing commitment to leading change in my profession.

1. **LAWS OF DISRUPTION:** As technology advances at an increasing pace, I recognize that technologies once invented never go back into the labs and that organizations and people are struggling to absorb the change. I will always answer the question, "Just because we can, should we?"

2. **ACCESS:** Advancing technologies create an opportunity to balance inequality across the country and around the world. I will always ask myself and my team if there is a way to help others with what we are building.

3. **PAY IT FORWARD:** I agree to "pay forward" whatever positive benefits I receive.
   - For every introduction I get, I will provide an introduction to another person.
   - For every hour of advice I receive, I will give an hour of advice to someone else.
   - For every risk someone takes with me, I will take a risk with a different person

4. **DISCOURSE:** I will take time to engage in the global debate on the role of professions and the impact of the change coming. I understand that as a leader of change I can be a positive and active voice in the debate. The moral issues of Artificial Intelligence and my profession are important and must include my active voice to help understand and advance.

5. **HUMILITY:** I will be transparent and honest. By acting in a spirit of humility I will never assume that I know more than someone else. I will also recognize the real challenges and pain that people feel. I serve them in all that is good and not good.

6. **RESEARCH:** I will dig into the issues that matter to those I serve. I won't give trivial answers to hard questions. My technology is part of a greater ecosystem.

7. **SHARING:** I will open myself to learning from others. I am eager to act to learn. I will share my knowledge in the spirit of the Creative Commons to help nurture learning in others.

8. **ROLE MODEL:** I will lead at times and follow at other times. Each person acts as a role model for everyone else. I will live these ideals and enforce them as a member of my community.

SIGNATURE _____ DATE _____

for human beings? To which class of society do we assign the gatekeeping function for the professional knowledge formerly held exclusively by human beings? With whom (or with what) does society entrust these decisions and interventions?

The authors argue that these discussion and debates need to happen sooner rather than later:

> ...we call for public debate on the moral issues arising from models for the production and distribution of practical expertise that do not directly involve profession or para-professionals. And we ask that this debate be held sooner rather than later, before our machines become much more capable...If we imagine a future in which much practical expertise can be made available online, it is less obvious that the profession, or indeed anyone, should be entitled to act as its gatekeepers.[114]

Perhaps the most important role that current leaders of professions must play is informing up-and-coming professionals of these issues and inviting them into the debate. Trying to re-invent the professions from within the ranks of the professional business model is virtually impossible. The economic rent paid to those with expertise and exclusive ownership of intellectual capital is very high. There is a huge disincentive for anyone making 10 to 100 times the average wage of the population to break apart the model. My mother's favourite expression was "Pull up the ladder, Jack, I'm on board," which she would use any time her children were inconsiderate toward others, be it not passing the potatoes at the dinner table or looking away from the disadvantaged on our walks. This applies to a subset of the professionals with whom I have worked over the years. The classic grind of overwork and low pay that characterizes the typical journey of the apprentice is followed by an extraordinary pay-off when he or she reaches a leadership role. It is simply the way the model has worked for a century or more.

Changing this model requires corporate bravery and policy leadership. These are often found in the governance bodies that oversee a particular profession—for example, the American Medical Association. These agencies can and must provide the platform for discussing the issues of a new social contract with the professions.

As with other roles in our life, we need a new social contract for the professions, both as a commitment to their profession and to their clients. On page 196 is a sample.

TEACHERS

It should be clear by this point how important education and new skills development will be in the future and just how much stress and change the education system in the developed and developing world will be under in the next 15 years.

I feel for Mervis and Olivia. Their cohorts will be greatly challenged by having to excel in a paradigm of learning that is fundamentally changing, just as they prepare for advanced education and their first jobs. At the same time, they are constantly being told that they will have to engage in lifelong learning. In short, they are "taught in the old way, underprepared for the new."

They must look back on their parents' generations and pine for a simpler model.

But all is not lost. There is much momentum across the world in creating the new world of teaching and education. We met Saul Khan and his extraordinary Khan Academy earlier.

Long promised but only now being delivered, the

technology that enables lifelong learning and the ability to track progress against personalized goals is becoming real. The amount of investment capital in learning technology companies has exploded. One estimate saw the overall private capital funding jump from $180 million in 2009 to over $1.3 billion in private capital in 2015.

The optimists are vocal and the prize is huge. As one leader in the teaching profession grandly professed,

> *We stand on the cusp of a great opportunity to end generations of educational discrimination and inequity, finally to fulfill the promises of our democratic republic. I believe the noblest teachers, students, and leaders of 2030 will be remembered by future generations as those who surged over the barriers to true public education and a fully realized teaching profession—while myopic former gatekeepers staggered to the sidelines of history.*[115]

Lofty prose, but the call to action in this 2011 article and many others is clear to me: Education is the archetypal "linear" organization I spoke about in Law #3 of Gibson's Laws of Disruption. We own it with our tax dollars; we own it because it owns the lion's share of our children's minds and development during their most formative times. It is at the VERY tip of the spear and represents the front line of the battle to confront change. For Mervis, it represents life itself. There is no bigger issue to confront.

Where to start?

In *Teaching 2030: What We Must Do for Our Students and Our Public Schools—Now and in the Future*, Barnett Barry and his co-authors present four emergent realties that could help schools and the teaching profession move forward in the next 20 years. Their themes bear repeating and understanding. I summarize their important themes below:[116]

Emergent Reality 1: *Digital Technology Comes of Age*

The learning environment is being transformed. Digital tools allow students to learn 24/7 and to develop in-demand skills. Many of the same tools allow teachers to learn from each other anywhere, at any time. And—as importantly—such technologies help teachers share more accurate data about student learning with policymakers and the public, boosting accountability.

Emergent Reality 2: *Physical and Digital Worlds Converge*

Expert teachers will create "seamless connections between learning in cyberspace and in brick-and-mortar schools."[117] These educators know how to reach the "iGeneration" student and how to serve as community organizers. Even as online learning explodes, an unstable economy and growing socioeconomic divides will require that teacher-leaders build strong school-community partnerships, connecting students and their families with a wide range of integrated services.

Emergent Reality 3: *Teaching Becomes a Matrix Profession*

New professional pathways will allow teachers with different skills and career trajectories to maximize their strengths. Educators will operate within career matrices, not old-school hierarchical ladders. Schools will employ an intricate array of specialists and generalists. Some will teach for only a few years. Some may teach solely or partially in online settings. Schools (even high-need schools) will be led by those who are committed to teaching for the long haul. Every school will be anchored by a core group of accomplished teachers who know deeply the students and families they serve.

Emergent Reality 4: *Entrepreneurial Thinking Starts Early*

The *Teaching 2030* authors predict the need for 600,000 "teacherpreneurs."[118] These are effective teachers who continue to work with students regularly, but also have the time, supports, and rewards necessary to apply their expertise in other ways. For example, teacherpreneurs may mentor new teachers, design new instructional programs based on gaming technologies, orchestrate community partnerships, or advance new policies and practices. Teacherpreneurs will be the "highest-paid anybodies"[119] in a school district—and their roles will finally blur the lines of distinction between those who teach in schools and those who lead.

In summary, the mantra of "learning together and from each other"—peer to peer learning—will dominate. Teachers will become more like facilitators of communities built around shared learning and aspiration. The role of mentors will increase; as part of a new social contract of the professions discussed in the previous section, bringing professional experience to technically literate students will gain importance.

The work that we are doing in our local communities here in Alberta mirrors and supports these and other trends. And it is happening all across the world.[120]

In the developing world, there is some promise but it is still evolving. The socio-economic and demographic challenges we talked about in our discussion of the developing world generally and the narrative of Mervis' journey specifically point to an uncertain future. The connected planet and the tip of our spear will help. But, as we have discussed many times, we will need to establish social contracts with those who will lead the revolution before the momentum truly shifts.

In summary, my heart and head are deeply rooting for success and change in the way we teach and learn. Similar to our other discussion, I am certain that it begins with a new commitment and an explicit social contract.

For those of you who are teachers or students, or simply concerned citizens, here are some thoughts:

- The current generation of kids like Mervis and Olivia are going to need help.

- We need to understand the transition that they will face and recognize that many of "linear" school systems will struggle to find resources, capacity, talent and time to adapt.

- Failure to adapt across the socioeconomic spectrum is a Tip of the Spear problem; it will be an existential issue for some of our communities.

- Start simple. At the earliest age, bring into your classroom real people from the "near future": real entrepreneurs and innovators who are inventing the future.

- Know and challenge the scientific evidence about how an "always on" world is affecting our children's minds.

As a teacher, understand that your field will be undergoing deep change over the next decade. Keep up and demand that there be a deep and ongoing collaboration between the governing bodies (trustees, etc.), curriculum designers, students and the world of innovation.

On page 203 is a social contract template that embodies many of these thoughts for a teacher. There could be similar

SOCIAL CONTRACT

TEACHER

This agreement describes what we mean to be an active and informed citizen in the world of change and innovation. It defines what we allow and pursue versus what we resist and avoid. It is a litmus test for the quality of our decisions, actions and ways of being together. By signing this agreement, I agree to live these principles as part of my ongoing commitment to leading change in my profession.

1. **LIFELONG LEARNING:** I am committed to advancing the means and ways of lifelong learning: in myself, my colleagues and my students.

2. **ACCESS:** Advancing technologies create an opportunity to balance inequality across the country and around the world. I will seek ways to provide access to the best tools available and actively call upon technology partners to support the challenges of the underfunded public systems around us.

3. **PAY IT FORWARD:** I agree to "pay forward" whatever positive benefits I receive.
   - For every introduction I get, I will provide an introduction to another person.
   - For every hour of advice I receive, I will give an hour of advice to someone else.
   - For every risk someone takes with me, I will take a risk with a different person.

4. **DISCOURSE:** I will take time to engage in the global debate on the role of teachers and education and understand the role I can play. I recognize that as a leader of change I can be a positive and active voice in the debate. The issues of inequality in the public systems of education are moral ones and my voice is needed to help how and where I can.

5. **HUMILITY:** I will be transparent and honest. By acting in a spirit of humility I will never assume that I know more than someone else. I will also recognize the real challenges and pain that people feel. I serve them in all that is good and not good.

6. **LEADERS AND ENTREPRENEURS:** I will be a facilitator and clearing house of the best and brightest in the world of entrepreneurs, new critical thinking and the skills that are essential in the new world of learning that will enable our children to solve the hardest problems in the world today.

7. **SHARING:** I will open myself to learning from others. I am eager to act to learn. I will share my knowledge in the spirit of the Creative Commons to help nurture learning in others.

8. **ROLE MODEL:** I will lead at times and follow at other times. Each person acts as a role model for everyone else. I will live these ideals and enforce them as a member of my community.

SIGNATURE _____ DATE _____

templates for a school district or for a university. Or perhaps, a social contract between Olivia, Mervis and their parents that commits to participating deeply and thoughtful in the coming decade of learning.

SUMMARY

The preceding examples are but a small snapshot of the challenges facing each of these roles at the tip of the spear. There are many more. We need to hold discussions on the new social contracts that will define and shape the emerging democratic and civil roles in developing Africa. We haven't even touched the evolving relationships in China and other Asian countries.

Each of these discussions represent independent studies and books in their own right. The research needed is underway and is deep and broad. I have included many of the research areas links at *theSpear.co* and I continue to follow them in earnest.

For those I have included, I have given starting points for you share the principles of how you and those in your specific circle wish to behave; by being explicit, you are able to begin the process of building a culture of trust. If the Rainforest movement has taught me anything it is this: individuals work most effectively when there is a shared and explicit culture of trust. If we are going to share new ideas with those who potentially do not agree with us, it is best that we make an honest and authentic attempt to trust and respect the other's thoughts and opinions.

In summary, I recognize that you can choose to be involved or not. But I will argue that, at a bare minimum, you have a responsibility to your family—whether nuclear or extended—and especially to those who are or will be affected by disruption and change.

For your family, simple involvement is having an honest discussion with your children about what the near future holds for them. Share articles that go beyond the trivial and look more deeply into the social, economic and political issues that these changes will bring. Put down your mobile device completely at least one full day a week and have a conversation with your family; read an article longer than five pages. But whatever you do, please don't buy into the fear mongering.

There is much to be optimistic about. The best of what makes us human—and we all have examples of this in our lives—comes through when we need it most. I cannot believe that all of the goodness of the human mind and spirit could invent the technology and the means to solve all of the existential problems of our world and—in the end—lose it all through squabbles that scratch the basest of our human cultural DNA. We have the control and we have the ability to make the leap. The tip of the spear might poke at but it can never be allowed to pierce our fragile hearts.

From the Fourth Industrial Revolution conclusion,

> ...neither technology nor the disruption that comes with it is an exogenous force over which humans have no control. All of us are responsible for guiding its evolution, in the decisions we make on a daily basis as citizens, consumers, and investors. We should thus grasp the opportunity and power we have to shape the Fourth Industrial Revolution and direct it toward a future that reflects our common objectives and values.[121]

It starts—as it always does—with some basic truths.

As Douglas Rushkoff, the American media theorist,

writer, columnist, lecturer, graphic novelist, and documentarian, beautifully notes in an interview from Singularity Hub called *Staying Human in the Machine Age: An Interview With Douglas Rushkoff*:

> *The more you do at a human scale—and I mean a human scale directly with other people, not just Skyping with other people, but really there in person—the more likely you are to be able to transform the landscape so it no longer favors just corporations and other abstract entities but you and your loved ones and your community.*[122]

Before we leave our Social Contract discussion, I need to call out directly to business leaders. We are increasingly looking to the social responsibility of our corporate leaders as they grow increasingly powerful, global and influential in our daily lives.

My call is not "to be more socially responsible." I have made the assumption that anyone leading an organization of any size recognizes that they have a role in the communities in which they sell their products and services. Rather, I call out to organizations to begin filling in the blank spaces that our schooling and other experiences have left us: The Skills of the New Business Leader.

chapter 19

SKILLS OF THE NEW BUSINESS LEADER

I have a separated this chapter because I think the challenge to our business leaders is particularly daunting. Every single leader in the world has to understand and contemplate to what degree their businesses are in the strategic cross hairs of change and disruption. All need to ask the basic question: "Do I understand enough of the coming disruption to prepare all of my stakeholders for the changes, many unanticipated, that will occur?" They need to ask, "In what way will our product or service still make sense in a resource constrained world?" Or even better, "What are the ways you can make your team be exponential and different?"

It is this last point that I believe requires significantly more digging into because it is at this front that the battle for new thinking will be fought the hardest.

Collective Brilliance—A New Social Contract for Leaders

If you have read Walter Isaacson's excellent biography of Apple CEO Steve Jobs, you will know that one of the central themes of the book is the nature of business leadership. The book juxtaposes Jobs' "corporate misogyny" against the failures and ultimate global success of Apple and Pixar and raises a fundamental question about leadership: "To make extraordinary things happen, does a leader have to be an a—hole?"

Or, put less colloquially: "Are great strides achievable only through huge and often 'extra-human' sacrifice?"

The stories of Mark Zuckerberg's leadership style at Facebook are legendary. Similarly, at Microsoft, a meeting with Bill Gates would often end with "Are you just ignorant or stupid?"[123] So, too, are the latest scandals from Uber and the crushing defeat of the infamous "pharma bro" Martin Shkreli.[124]

Is that what leadership has become in technology companies? You are either a God or you suck; your product succeeds quickly or it fails fast. Binary. Black. White.

While in rare cases these leadership approaches can generate huge wins, they are also hugely divisive and create enormous costs for employees—and society. We have spent an entire chapter discussing the increasing evidence of the heavy impact of social and economic inequality. But more tactically and in terms of the future, these examples are also toxic models to emulate for the emerging leaders of the entrepreneurial start-up, a medium-sized company, or large organizations.

The real truth is that leading a modern technology company is a far more complex calculus and requires a new and fully revised social contract and approach.

The obvious questions to me and many others reading the Isaacson book are:

- Does creating true innovation and shifts in paradigm—something Steve Jobs did at least three times in his career—require a brutality and single-mindedness that runs counter to norms of civil behaviour and management motivation?

- Could a collaborative and more nurturing leader with equal brilliance have delivered the same results?

- Are Elon Musk, Steve Jobs, Bill Gates, Mark Zuckerberg the leadership archetypes for the next decades?

These fascinating questions speak to the heart of leadership and innovation for the emerging companies of the world.

My hypothesis is that, as we become increasingly comfortable and capable of deep and meaningful collaboration across cultures and ecosystems; as our ability to reach out and touch and process new ideas and concepts from our global customers, partners and stakeholders increases; and—most importantly—as more of our leaders emerge from the generational talent pool that Don Tapscott calls "Grown Up Digital" (the globally connected Olivia and Mervis, for example), the idea of what makes a successful leader is profoundly changing.

Put simply, I think we have become too smart and too connected to tolerate simplistic, ego-driven command and control hierarchies. I think our teams, staff, employees and boards expect better and—more importantly—are easily able to absorb and thrive in a more complex, socially connected ecosystem. They will demand a new type of "enlightened" leadership. As new tools and cultures evolve over the next decade, I believe what will emerge is a type of "ecosystem" leadership that will be central to a new leadership style I call *Collective Brilliance*. It fits perfectly with the ideals explored earlier on the new Social Contract and it is a gauntlet thrown down to organizations of all cultures, of all sizes and all types to appreciate the role they play in shaping the attitudes of new leaders and ultimately the environment of the people with whom they work.

Let me give you an example of this new type of leadership

and use it to identify the four principles of Collective Brilliance.

Collective Brilliance in Action: *Creating a Ballet*

The creation of a ballet from concept to creation to opening night is akin to what happens in a technology start-up. I am proud to have two adult children, twins, who are talented professional ballet dancers. I have served on the board of a $15 million ballet company and I have watched the extraordinary journey of how the dancers, the choreographer and the production teams come together as a new ballet is created.

In short, I have seen "how the sausage is made." And it is a true example of "collective brilliance" if done right.

PRINCIPLE ONE: INITIATE WITH PASSION

Lead through Inspiration and Truth

Choreographers are visionaries in the truest sense of the word: They SEE something that has NEVER existed before. They feel it with a consuming passion; they imagine movement and music meshing together; a particular piece will consume and drive them.

The selling of this vision is single-minded and single-handed—it is not a very collaborative effort. It is cajoling, selling and begging. It is about getting people who DO NOT SEE the vision to SEE it. It's about translating the abstract to something that "mere mortals" can comprehend. Why? Simple: Art is expensive and it needs to be sold to people with money.

Sound familiar? The best artistic directors and

choreographers pitching a new work would rival the best of anything Silicon Valley entrepreneurs—past or present—have to offer. I know. I have seen both.

More importantly, great artistic directors and choreographers know that they themselves cannot dance the dance. Their passion must inspire the artists. When the lights go down and the curtain comes up, the vision and passion must translate into action, interaction and movement. Partially script, partially magic, it must be made personal by every individual dancer and it must come together seamlessly as a group or duet or ensemble. The singular, passionate vision becomes the troupe's mission through inspiration and perspiration.

> The mediocre teacher tells. The good teacher explains. The superior teacher demonstrates. The great teacher inspires.
> *William A. Ward*

Why This Matters Now:

- We are living in a time desperate for visionaries and leaders who can make real change happen; leadership matters now more than ever.

- The newest generation of working professionals is increasingly cynical and less trusting of authority.

- Vision and inspiration are antidotes to cynicism and mistrust.

Business Leadership Summary:

- Greatness and vision do indeed come from passion, but to move it forward the great passionate leader needs to inspire.

- Spend the time and effort required to ensure everyone on the team can internalize the vision.

- Translate the passion into inspiration; unbridled passion wears people out. Finding out what inspires teams and individuals is hard but it's a must if the project, company or product is to move forward.

PRINCIPLE TWO: ACCELERATE WITH COLLABORATION AND COLLECTIVE WISDOM

Once green-lighted, the ballet creation process shifts to a fast-paced, collaborative, tough and singled-minded journey implemented by consummate professionals.

It all develops over many months using a fascinating hive process. But trust me, the collaborative leadership model isn't weak and it is not for the faint of heart. If you have ever seen a choreographer pull his or her hair out during the difficult transition from concept to movement and some form of coherency, you will appreciate that there is nothing soft about it. It can be ruthless. It shouts, it swears, it cajoles, and it can indeed be "I'll know it when I see it." "AH-HAH !!!", shouts the director, "That's IT!!!" (Now do it again and again 400 hundred more times!)

The production side of a ballet is a global collaborative effort. It is a different kind of choreography, delivered mostly by virtual, geographically dispersed global teams.

This collective intelligence collaborates in every imaginable way—some effective and efficient; others fragmented and frustrating.

The clear message is that the original passion morphing into an inspiration-led effort is as collaborative as anything I have ever witnessed in business. Driven by a fixed date called "Opening Night," this inspiration is often characterized as sheer terror, but more often it is managed and motivated through the continued inspiration of the creative leader.

Why Now?

- Solutions to local or global problems are increasingly becoming more than what just one person or one company can solve alone.

- The social enterprise is becoming populated by connected and aware employees who are comfortable with and will demand access to the tools and cultures that embrace community and collaboration.

Business Leadership Summary

- Tap into collective intelligence and provide the systems and process to enable dispersed teams to engage and work consistently toward the goal.

- Break down the physical, cultural and technical barriers that make distributed work so challenging.

- Invest in tools and processes that celebrate and enable collaborative decision making and collective intelligence.

PRINCIPLE THREE: DEMAND PROFESSIONALISM AND PERSPIRATION

Have you ever seen a ballet class or witnessed young boys and girls aspiring to dance in one of their many daily sessions in the multi-mirrored hell called the ballet studio?

I have often said that I would love to take a few of the so-called professionals/creatives I have met in the business world and force them to watch a ballet class to understand what it means to be a real professional: Focused work; repetition; hours and hours critically staring at your "instrument" in the mirror; challenged for more at every turn; class, rehearsal, repeat.... To even get to be considered a professional takes all of Gladwell's 10,000 hours and is a study in discipline, simplicity, focus and pain. I would challenge many of the overpaid so-called professionals in the business world to hold their professionalism up to that standard.

The reason: To make magic happen on stage requires something a lot less than magic in the studio. This is the other side of dance that most people don't ever get to see. It is fundamentally about driving the choreography into the nether regions of the brain so that it becomes rote. Over and over it is repeated. Ten seconds in the ballet can be rehearsed for half a day.

It's because when that happens, the dancer can actually DANCE beyond the steps. They can move. They can deliver the emotional elements that take it from a series of pretty and athletic steps to art, emotion, passion and—hopefully—something that can cause the audience's hearts to collectively skip a beat. And it's the hardest thing I have ever witnessed. We hear about the same muscle memory in the most elite of our athletes and military special forces.

The best directors know that it is the underpaid and passionate professionals that deliver the goods. No matter the praise heaped on the show, at that moment on stage it is about the people who put everything on the line and made it happen. And if you have ever been back stage at such a performance you will know that there is always considerable drama and stress. But the customer never sees it. The show goes on. The team rallies, the petty dramas of the rehearsals of months' past disappear; it's all about the audience.

The bottom line: If you are in business, hire the best, demand the best and give them every opportunity to fine-tune their craft. There is no other way. In the ballet, the dancers and team constantly have to learn new choreography, sometimes for multiple performances at the same time. Individual dancers can be the star of the show once, supporting the next; they are consummate team players, fast learners and constantly keep their instruments tuned. Sound like a blueprint for the ideal team in a fast-paced tech start-up?

Why Now?

- New business building, design, development, marketing and implementation is a very complex calculus. It requires talent that can handle hard work, disciplined feedback and multiple objectives while managing their place in a complex group dynamic.

- Critical thinking is more and more rare but is essential for achieving breakthrough insights. Disciplined and trained professionals understand how to manage distraction and embrace the simple solution. As Einstein said, "Everything should be made as simple as possible; but not simpler."

Business Leadership Summary

- Great work is hard. It requires a level of discipline and professionalism that happens over many hours and years; hire professionals and demand professionalism in all aspects of the job.

- Understand that a chain is as strong as the weakest link and that the ultimate evidence of professionalism is in doing everything and more to ensure you hold up your social contract with the team.

- To be agile, a company's best people have to be spry while maintaining quality and professionalism.

PRINCIPLE FOUR: SUPPORT THROUGH ENGAGEMENT AND RECOGNITION

Dancers in an average ballet company (especially in North America) earn approximately 50% of the average median salary of the full-time workforce. They train longer than doctors to become professionals; they are often physically scarred and emotionally fragile—they live, after all, an insecure, contract-to-contract life and are one snapped ankle from the unemployment line.

So why do they do it? At the heart of the answer is the difference between intrinsic recognition and external rewards. As passionate professionals to whom our North American society has decided to pay subsistence wages, they are driven by something much deeper than money. The intrinsic rewards that come from driving to perfection, taking their physical and emotional selves beyond what they thought possible and being on stage to the sound of the audience appreciation are what motivates most dancers.

Studies have clearly shown that recognition based on human values—or "intrinsic recognition" is a far more powerful motivator then extrinsic reward-based systems which provide prizes or goodies as reward for certain behaviours or achievements. It's why companies such as KudosNow (*www.kudosnow.com*) have emerged in response. They have recognized that the best of us need much more than money to be motivated. Many corporate performance studies show a direct and positive correlation between employee engagement and the company's financial performance. Yet recent studies demonstrate that the majority of the workforce is disengaged. Lost productivity in the US alone is estimated at $370 billion annually.[125]

Would the average dancer want to earn more money? Absolutely. But that they do what they do, day in and day out over many years, suggests much more is going on. I have spent many hours with these special people and I can tell you that they can teach us about the balance between the simple pleasure of doing great work, extrinsic rewards and recognition. What they do has meaning and in my experience as a leader, in the absence of meaning no amount of money will retain the best.

Why Now?

- To the newest generation of professionals, meaning matters.

- Engagement with an inspired purpose is becoming a real requirement in attracting and retaining talent.

Business Leadership Summary

- Base your compensation and performance on a recognition system as well as reward systems.

- Understand that the best are professionals and professionalism needs engagement and recognition, not rules and reward.

SUMMARY

"Collective Brilliance" is alive every day in the creative process of creating captivating art such as a ballet. While it can be messy and suffer from the imperfections and drama of everyday life inside an organization of well-meaning people, it *works*: very rarely does the show not go on. Not every performance or vision is a four-star rave. But the art does survive and the process has much to teach us. The social contract is explicit and clear.

So, back to Steve Jobs. The net for me is that Steve (and Elon, Mark, and Bill) were and are very rare creatures whose skills as leaders will be greatly debated for a long time. I don't believe, however, that their models are ones to emulate.

I do think that today's leaders, especially in the crucible of technology and innovation, can deliver massive innovation and breakthrough performance if they recognize some of the simple truths of how professionals work, how to harness the collective, and how to translate passion into inspiration. Above all, great leaders need to build new social contracts and surround themselves with individuals who value intrinsic recognition over extrinsic reward, individuals who can be inspired and can inspire their audience/customer. Professionalism, perspiration, inspiration, collaboration and engagement are the new watchwords of leadership in the era of Collective Brilliance.

As you begin your journey in building new social contracts with your team, keep these principles in mind:

Collective Brilliance: Business Leaders' Social Contract

Summary

INITIATE WITH PASSION; LEAD THROUGH INSPIRATION

- Greatness and vision do indeed come from passion, but to move them forward, the passionate leader needs to inspire.
- Spend the time and effort required to ensure everyone on the team can internalize the vision.

ACCELERATE WITH COLLABORATION AND COLLECTIVE WISDOM

- The best and the brightest exist out there. The new leaders tap into collective intelligence and provide the systems and process to enable these dispersed team to engage and work consistently toward the goal.
- Break down the physical, cultural and technical barriers that make distributed work so challenging.

DEMAND PROFESSIONALISM AND PERSPIRATION

- Great work is hard. It requires a level of discipline and professionalism that develop over many hours and years; hire professionals and demand professionalism in all aspects of the job.
- To be agile requires the best people to be able to move quickly and maintain quality and professionalism.

SUPPORT THROUGH ENGAGEMENT AND RECOGNITION

- Base your compensation and performance on a recognition system versus a reward system.
- Understand that the best are professionals, and professionalism requires engagement and recognition—not rules and reward.

chapter 20

LEAD, FOLLOW, OR ...

Men go forth to marvel at the heights of mountains and the huge waves of the sea, the broad flow of the rivers, the vastness of the ocean, the orbits of the stars, and yet they neglect to marvel at themselves"

Saint Augustine of Hippo, Confessions

Thinking back to my connection between Ray Kurzweil and Donald Trump from earlier: on reflection, I think the juxtaposition of these two archetypes comes at a very important time. I am convinced that the forces of change are coming very quickly. The bigger story of Trump and the disenfranchised gives us early warning signs. The canary hasn't died yet, but after the upcoming U.S. 2018 election cycle runs its course, it will have emerged from the coal mine gasping and frightened about what lies below.

The journeys of Olivia and Mervis have allowed us a peek of what is in store for our children's future. The decisions we start making today—as parents, as leaders, as voters and simply as human beings—may or may not make their journey any easier. On a continental scale, the decisions that Africa must make with the developed world's assistance are urgent, large and uncertain.

The journey of Mervis and Olivia will, as we all know, be the sum of the decisions they make over a decade or more. It is up to them to a large degree, but what we do as a society will signal our intent. It will signal that we felt the tip of that spear poking at our ribs and that we chose to act.

I hope this book has started a conversation in your mind, in your heart and with the people you love and care about. Whatever your reaction, I wrote this intending to give you pause and to provide some tools to reflect on the many roles you play in your day-to-day life as the future unfolds before you.

It started from that proverbial gut feeling that something wasn't right. My decades in the technology industry gave me an unvarnished view into its benefits and perils. I have been growing increasingly worried, but I believe the story finishes with hope. I am cautiously optimistic.

Existential Crises

After all of this discussion, angst and conversation, however, it comes down to what it always comes down to: you and me. And we have plenty to do. We have three existential crises facing our planet right now that require us all to drop our pens, remove our blinders, shelve our old ideas, holster our weapons... and figure it out.

The first, inequality and the distribution of wealth resulting from exponential technology disruption, has been the subject of this book. It is perilous, it is happening now, and it needs some very specific attention. The journeys that our two hypothetical girls may take into this frightening future are very real. At a minimum, I ask you to imagine a 13-year-old you know. Think of her journey ahead. Think of the decisions we make and the actions we take and how they might alter her path. This is an existential crisis entirely of our own making and as such it is fully in our hands to help Olivia and Mervis cope and, indeed, survive. My hope in writing this book is that if we make progress in keeping the tip of the spear from piercing our human hearts, the rest will follow.

The two other existential crises also require immediate attention and focus. Human-caused climate change is similar to our tip of the spear conversation. The implications for our species are real and immediate. But if the hypotheses being tested by good science hold and the evidence fits, we have no time for long-term discussions. My sense from my own research is that we are well past the discussion stage and it is time for bold leadership. My deep hope is that the enormously difficult and contentious switch to alternative energies over the coming decades will set an example for a new partnership between the private sector and the public governance model that—if I read the economic and broad public policy debate correctly—could radically alter both for the better. But that is the subject for another day and another book. Our planet is asking for a new social contract. I am worried but hopeful that we can write it, soon.

Our third existential crisis is what others have called "the last invention of man"—Artificial Intelligence. I have separated AI from other technology disruption because it is at the same time the last technology we will build on our own and the defining platform from which we will debate truly global issues as a species. And much of this continues to happen in real time: Here is the latest (September 2017) from the leader of Russia, clearly reminding us what is at stake:

> "Artificial intelligence is the future, not only for Russia, but for all humankind," [says] Putin.... "It comes with colossal opportunities, but also threats that are difficult to predict. Whoever becomes the leader in this sphere will become the ruler of the world."[126]

Remember the movies where the aliens came to earth and forced all nations of the planet to somehow figure out what to do as a species? In my mind, AI is just like that.

It will go to the heart of the inequality discussion and it will go to the heart of how we decide to govern ourselves as a species. It embodies all of the three laws of disruption. It is clear from the research and massive movements of capital and resources into the AI arena that this genie is out of the bottle. And we have been warned.

At the end, as we discussed earlier, I have had the good fortune of watching a small social movement take hold and flourish. I can tell you from this experience that the power of ideas delivered in a culture of mutual understanding and trust can be exponential. It can match any change technology acceleration throws at us, and can spin the flywheels of linear organizations faster than they have ever spun before. The power of human minds brought together in a spirit of competitive trust is staggering.

The reason I used the journey of 13-year-old girls to tell this tale is that they are walking right into the middle of this. I spoke of the Great Social Experiment. I noted that Olivia and Mervis will be the generation that will cross Ray Kurzweil's singularity as they hit their 30s—if we make it that far. I believe they truly need our help but, just as important, we need to listen to and understand them. Olivia needs to grow into a world with a new kind of leadership that protects her, teaches her, guides her and sets her out a journey with the skills to adapt. Mervis needs the tent pegs of democracy, connectivity and education to be firmly anchored so that she can unleash her potential on the world, unencumbered by the experiments of the past century of progress that have given great wealth to Olivia but have brought inequality and lack to her.

My simple plea to the reader is to please stop waiting for

someone else to do something. As the Rainforest movement taught me, two things are foundational for large scale systemic change:

- Understand the power of creating a culture of trust.

- Understand the power of leadership and role models.

Think of all of the interactions you have had in your life that would have been different if you had actively encouraged trust. It manifests itself everywhere, doesn't it? The conversations you have online, the interaction with your neighbor and the work you do every day as a leader or a follower, as an independent or a member of a group.

I am reminded of the famous quote often attributed to Thomas Paine, General Patton and made famous by Lee Iacocca, the former CEO of Chrysler, in the turbulent automotive industry of the 70s: *"Lead, Follow or Get Out Of the Way."* As I reflect on our story, I am very tempted to use this as my concluding theme. It reflects a simple passion combined with a bit of passive aggression that (if I am really being true to what's going on in my head) I *want to* believe.

There has never been a more important time for leadership of our species as right now. But where are the leaders that we will follow? We have invented the tools of the gods and yet we bicker and waste our precious minutes, hours and years on the trivial. That is where the anger comes from and I simply want to say "Get out of the way."

But I am reminded by some extraordinary people I have met on this journey that change happens one human being at a time. I am further reminded that at a time when we have an incredible ability to share and connect across a city or even the planet, we can—instead of telling people to get out of the way—share and teach and collaborate.

But please, pay attention, give a damn and have some grit. Big waves are great when you can anticipate and ride them, but are horrifying when they toss you about and rip your bathing suit off.

And Finally

I happen to live in one of the most treasured and richest places on the planet and, while we are facing an important systemic challenge in our city, we have more or less won the location lottery. So while the changes that I have seen with our social contract may seem trivial when compared to the fundamental poverty and inequality of the developing world, they have taught me a valuable lesson about what is possible when good intentions and simple ideals are consistently applied.

I happen to believe that they are universal. And I happen to believe they are truly exponential.

Be bold. Aim your spear high and bring others along in your journey.

ENDNOTES

1. The library was called Memorial Park Library in the Beltline area of Calgary. Truly a lovely place. Robustly constructed and richly detailed, the Memorial Park Library is a fine representative of Edwardian Classicism, a pre-First World War architectural style related to the French Beaux-Arts style. It was built in 1912.

    calgarylibrary.ca/locations/MPARK/

2. Malawi. (2017, November 3). In Wikipedia, The Free Encyclopedia. Retrieved on 2017-11-09, from *en.wikipedia.org/w/index.php?title=Malawi&oldid=808595732*

    I was careful in my selection of the country and setting for my fictional Mervis. I have not travelled to Malawi and so the character and her location were perhaps the hardest aspect of the book. I was highly sensitive to not trivializing nor simplifying a complex story arc. In addition, in all of my readings and research I have been very careful (and I have been reminded) not to "broad brush" the African continent. It is as diverse a place anywhere on the planet.

3. *Girl Up | uniting girls to change the world | United Nations Foundation.* (n.d.). Retrieved from *girlup.org/#sthash.bCAMTIo2.dpbs* Retrieved on 2016-11-25.

    From the Girl Up Web Site:

    Girl Up, the United Nations Foundation's adolescent girl campaign, engages girls to stand up for girls, empowering each other and changing our world. Led by a global community of passionate advocates changing policies and raising funds to support United Nations programs, our efforts help the hardest to reach girls living in places where it is hardest to be a girl.

4. Ayemoba, A. (2016, March 3). *Hjörtur Smárason: African cities take the lead in developing the future City.* Retrieved from *africabusinesscommunities.com/features/hj%C3%B6rtur-sm%C3%A1rason-african-cities-take-the-lead-in-developing-the-future-city.html*

5   The quote is most often attributed to William Gibson though it has been widely discussed that he never actually wrote it down. From the website Quote Investigator—*quoteinvestigator. com/2012/01/24/future-has-arrived/*.

They note: *In conclusion, the evidence is strong that William Gibson used this expression and* **QI** *believes that he created it. However, the precise wording varies and the earliest citations do not appear in Gibson's writings. In 1992 the journalist Scott Rosenberg writing in the San Francisco Examiner credited William Gibson with the maxim as mentioned previously in this article. This is the earliest citation known to* **QI** *[WGSF]:*

As William Gibson put it: "The future has arrived—it's just not evenly distributed yet."

6   Bulkeley, William (1989-06-23). "Kurzweil Applied Intelligence, Inc". The Wall Street Journal. p. A3A. Retrieved on 2017-01-25.

7   Pfeiffer, Eric (1998-04-06) "Start Up". Forbes. Retrieved on 2016-11-25.

8   "Who Made America?". PBS. Retrieved 2017-09-11.

9   I encourage you to read *Age of Spiritual Machines: When Computers Exceed Human Intelligence* and *The Singularity Is Near: When Humans Transcend Biology* by Ray Kurzweil. They are deep, philosophical and very important books. But what is truly inspiring is when they were written: *Age of Spiritual Machines* was written 18 years ago; *The Singularity*, 12 years ago. This truly puts Ray at the edge of what's happening. We need to pay attention if he writes something next!

10 The dividing of America. (2016, July 16). *Economist*.

11 The writing of this book started during the rise of Donald Trump and the Brexit vote. This has made for an interesting author journey. In fact, the election of Mr. Trump happened while I was in London, UK writing the largest section of the material. Digesting the implications of the election changed the urgency and the focus of the book. It made it very hard to not take a darker turn and it was only the work of a number of authors and conversations that made this a "cautiously optimistic" story.

12 "Moore's law." (2017, November 10). In *Wikipedia, The Free Encyclopedia*. Retrieved 23:24, November 12, 2017, from *en.wikipedia.org/w/index.php?title=Moore%27s_law&oldid=809657628*

(From Wikipedia) Moore's law is the observation that the number of transistors in a dense integrated circuit doubles approximately every two years. The observation is named after Gordon Moore, the co-founder of Fairchild Semiconductor and Intel, whose 1965 paper described a doubling every year in the number of components per integrated circuit, and projected this rate of growth would continue for at least another decade In 1975, looking forward to the next decade, he revised the forecast to doubling every two years. The period is often quoted as 18 months because of Intel executive David House, who predicted that chip performance would double every 18 months (being a combination of the effect of more transistors and the transistors being faster).

Moore's prediction proved accurate for several decades, and has been used in the semiconductor industry to guide long-term planning and to set targets for research and development.[10] Advancements in digital electronics are strongly linked to Moore's law: quality-adjusted microprocessor prices, memory capacity, sensors and even the number and size of pixels in digital cameras. Digital electronics has contributed to world economic growth in the late twentieth and early twenty-first centuries Moore's law describes a driving force of technological and social change, productivity, and economic growth.

13 Image Source: Ray Kurzweil: The Singularity is Near: When Humans Transcend Biology",P.67, The Viking Press, 2006

14 Simon, W. G. (2013). *Transistor Count and Moore's Law—2008—Linear II* [Graph]. Retrieved from *en.wikipedia.org/wiki/File:Transistor_Count_and_Moore's_Law_-_2008_-_Linear_II.png*

15 Andreessen Horowitz. (2015, November 18). *Software is eating bio*. Retrieved from *www.slideshare.net/a16z/software-is-eating-bio*

Actually, anything written by the Andreessen team is worth reading. Subscribing to their blogs and content is probably the single best source of material you will find on future trends and business.

16 Diamandis, P. H. (2015). *Abundance: The future is better than you think*. New York, MT: Free Press.

17 Kurzweil, R. (2017, July 7). *Ray Kurzweil's Mind-Boggling Predictions for the Next 25 Years*. Retrieved from *singularityhub.com/2015/01/26/ray-kurzweils-mind-boggling-predictions-for-the-next—25-years/*

18 Frank, A. (2017, May 19). *The World Depends on Technology No One Understands*. Retrieved from *singularityhub. com/2016/07/17/the-world-will-soon-depend-on-technology-no-one-understands/*

19 Diamandis, P. H., & Kotler, S. (2016). Exponential technology: The democratization of the power to change the world. In *Bold: How to go big, achieve success, and impact the world* (p. 106). New York, NY: Simon & Schuster.

20 *Experts—Singularity Hub*. (n.d.). Retrieved November 12, 2017, from *singularityhub.com/experts/*

21 Cann, O. (2017, June 26). *These are the top 10 emerging technologies of 2017*. Retrieved from *www.weforum.org/agenda/2017/06/these-are-the-top–10–emerging-technologies-of–2017/*

22 Dixon, C. (2016, February 21). *What's Next in Computing? ? Software Is Eating the World? Medium*. Retrieved from *medium.com/software-is-eating-the-world/what-s-next-in-computing-e54b870b80cc#.syetjrj2g*

23 Urban, T. (2015, January 22). *The Artificial Intelligence Revolution: Part 1—Wait But Why*. Retrieved from *waitbutwhy.com/2015/01/artificial-intelligence-revolution–1.html*

24 Tapscott, D., & Tapscott, A. (2016). *Blockchain revolution: How the technology behind Bitcoin is changing money, business and the world*. Toronto, Canada: Penguin Canada.

25 Diamandis, P. (2016, February 28). *Drones & Technology Convergence* [Web log post]. Retrieved from *peterdiamandis.tumblr.com/post/140156221733/drones-technology-convergence*

26 Kurzweil, R. (2001, March 7). *The Law of Accelerating Returns | KurzweilAI*. Retrieved from *www.kurzweilai.net/the-law-of-accelerating-returns*

27 *Computer pioneers—Herbert A. Simon*. (2016). Retrieved from *history.computer.org/pioneers/simon.html*

28 Stewart, D. (1994). *Q&A: Herbert Simon on the Future of Artificial Intelligence*. Retrieved from *www.omnimagazine.com/archives/interviews/simon/index.html*

29 *Computer Pioneers—Marvin Lee Minsky*. (2016). Retrieved from *history.computer.org/pioneers/minsky.html*

30 Allen, F. (2001, March 1). *The myth of artificial intelligence | AMERICAN HERITAGE*. Retrieved from *www.americanheritage.com/content/myth-artificial-intelligence*

31 Brynjolfsson, E., & McAfee, A. (2017, July 18). *The business of artificial intelligence*. Retrieved from *hbr.org/cover-story/2017/07/the-business-of-artificial-intelligence*

32 The term "open source" refers to something people can modify and share because its design is publicly accessible.

The term originated in the context of software development to designate a specific approach to creating computer programs. Today, however, "open source" designates a broader set of values—what we call *"the open source way."* Open source projects, products, or initiatives embrace and celebrate principles of open exchange, collaborative participation, rapid prototyping, transparency, meritocracy, and community-oriented development.

From : *What is open source?*. (n.d.). Retrieved November 12, 2017, from *opensource.com/resources/what-open-source*

33 Philip Snowden, 1st Viscount Snowden. (2017, May 22). *Wikiquote,* . Retrieved 22:24, November 12, 2017 from *en.wikiquote.org/w/index.php?title=Philip_Snowden,_1st_Viscount_Snowden&oldid=2254818*.

34 Lewis, N. (2015, December 1). How sensitive is global temperature to cumulative CO_2 emissions? Retrieved from *judithcurry.com/2015/11/30/how-sensitive-is-global-temperature-to-cumulative-co2-emissions/*

Extract: Global mean surface temperature increase as a function of cumulative total global CO_2 emissions from various lines of evidence. Multi-model results from a hierarchy of climate-carbon cycle models for each RCP until 2100 are shown with coloured lines and decadal means (dots). Some decadal means are indicated for clarity (e.g., 2050 indicating the decade 2041–2050). Model results over the historical period (1860–2010) are indicated in black. The coloured plume illustrates the multi-model spread over the four RCP scenarios and fades with the decreasing number of available models in RCP8.5. The multi-model mean and range simulated by CMIP5 models, forced by a CO_2 increase of 1% per year (1% per year CO_2 simulations), is given by the thin black line and grey area. For a specific amount of cumulative CO_2 emissions, the 1% per year CO_2 simulations exhibit lower warming than those driven by RCPs, which include additional non-CO_2 drivers. All values are given relative to the 1861–1880 base period. Decadal averages are connected by straight lines

35 Rochford, J. (2013). *Human population growth (AD1—2013)* [Graph]. Retrieved from *sites.psu.edu/colecivic/2017/01/12/an-exponentially-growing-concern-overpopulation/*

36 Pyne, S. J. (n.d.). How humans made fire, and fire made us human—Stephen J Pyne | Aeon Essays. Retrieved December 5, 2017, from *aeon.co/essays/how-humans-made-fire-and-fire-made-us-human*

37 Nicolaus Copernicus. (2017, November 28). In *Wikipedia, The Free Encyclopedia*. Retrieved 20:31, December 5, 2017, from *en.wikipedia.org/w/index.php?title=Nicolaus_Copernicus&oldid=812626968*

38 www.icanw.org/the-facts/the-nuclear-age/

The entire table is reprinted here for effect. It is an extraordinary journey that mankind has undertaken on one technology.

TABLE 1: NUCLEAR WEAPONS TIMELINE: ICAN WEBSITE

August 1942	Manhattan Project established in US	The US sets up the Manhattan Project to develop the first nuclear weapon. It eventually employs more than 130,000 people and costs US$2 billion ($25 billion in 2012 dollars).
16 July 1945	US conducts first ever nuclear test	The US government tests its first nuclear weapon, code-named "Trinity", in New Mexico. Its yield equals 20,000 tonnes of TNT. The date of the test marks the beginning of the nuclear age.
6 August 1945	US drops atomic bomb on Hiroshima	The US detonates a uranium bomb over the Japanese city of Hiroshima, killing more than 140,000 people within months. Many more later die from radiation-related illnesses.

9 August 1945	A second bomb is dropped on Nagasaki	The US explodes a plutonium bomb over Nagasaki. An estimated 74,000 people die by the end of 1945. Little can be done to ease the suffering of the victims who survive the blast.
24 January 1946	UN calls for elimination of atomic weapons	In its first resolution, the UN General Assembly calls for the complete elimination of nuclear weapons and sets up a commission to deal with the problem of the atomic discovery.
29 August 1949	Soviet Union tests its first nuclear bomb	The Soviet Union explodes a nuclear weapon code-named "First Lightning" in Semipalatinsk, Kazakhstan. It becomes the second nation to develop and successfully test a nuclear device.
3 October 1952	UK tests nuclear weapon in Australia	The UK conducts its first nuclear test at Montebello Islands off the coast of Western Australia. It later conducts a series of tests at Maralinga and Emu Fields in South Australia.
9 July 1955	Russell–Einstein manifesto issued	Bertrand Russell, Albert Einstein and other leading scientists issue a manifesto warning of the dangers of nuclear war and urging all governments to resolve disputes peacefully.
17 February 1958	UK disarmament campaign formed	The Campaign for Nuclear Disarmament in the UK holds its first meeting. Its iconic emblem becomes one of the most widely recognized symbols in the world.
1 December 1959	Nuclear tests banned in Antarctica	The Antarctic Treaty opens for signature. It establishes that "any nuclear explosion in Antarctica and the disposal there of radioactive waste material shall be prohibited"

13 February 1960	France tests its first nuclear weapon	France explodes its first atomic bomb in the Sahara Desert. It has a yield of 60–70 kilotons. It later moves its nuclear tests to the South Pacific. These continue up until 1996.
30 October 1961	Largest ever bomb test conducted	The Soviet Union explodes the most powerful bomb ever: a 58-megaton atmospheric nuclear weapon, nicknamed the "Tsar Bomba", over Novaya Zemlya off northern Russia.
16–29 October 1962	Cuban Missile Crisis occurs	A tense stand-off begins when the US discovers Soviet missiles in Cuba. The US blockades Cuba for 13 days. The crisis brings the US and Soviet Union to the brink of nuclear war.
5 August 1963	Partial Test Ban Treaty opens for signature	A treaty banning nuclear testing in the atmosphere, outer space and under water is signed in Moscow, following large demonstrations in Europe and America against nuclear testing.
16 October 1964	China conducts its first nuclear test	China explodes its first atomic bomb at the Lop Nor testing site in Sinkiang Province. In total, China conducts 23 atmospheric tests and 22 underground tests at the site.
14 February 1967	Latin America becomes nuclear-free	A treaty prohibiting nuclear weapons in Latin America, the Treaty of Tlatelolco, is signed at Mexico City. Parties agree not to manufacture, test or acquire nuclear weapons.
1 July 1968	Non-Proliferation Treaty is signed	Under the Non-Proliferation Treaty, non-nuclear-weapon states agree never to acquire nuclear weapons, and the nuclear-weapon states make a legal undertaking to disarm.

18 May 1974	India conducts first nuclear test	India conducts an underground nuclear test at Pokharan in the Rajasthen desert, codenamed the "Smiling Buddha". The government falsely claims it is a peaceful nuclear test.
22 September 1979	Nuclear explosion in Indian Ocean	A nuclear test explosion occurs over the South Indian Ocean off the Cape of Good Hope. It is thought to have been conducted by South Africa with the assistance of Israel.
12 June 1982	A million people rally for disarmament	One million people gather in New York City's Central Park in support of the Second United Nations Special Session on Disarmament. It is the largest anti-war demonstration in history.
10 July 1985	Rainbow Warrior ship destroyed	The Greenpeace ship Rainbow Warrior is destroyed in New Zealand on its way to the Murorua Atoll to protest French nuclear tests. New Zealand later enacts nuclear-free legislation.
6 August 1985	South Pacific becomes nuclear-free	The South Pacific Nuclear Free Zone Treaty is signed at Rarotonga in the Cook Islands. The treaty prohibits the manufacturing, stationing or testing of nuclear weapons within the area.
10 Dec. 1985	Anti-nuclear doctors win Nobel	The International Physicians for the Prevention of Nuclear War receives the Nobel Peace Prize for its efforts to bridge the cold war divide by focusing on the human costs of nuclear war.
30 Sept. 1986	Israel's nuclear programme revealed	The Sunday Times publishes information supplied by Israeli nuclear technician Mordechai Vanunu, which leads experts to conclude that Israel may have up to 200 nuclear weapons.

11–12 October 1986	US and Soviet leaders discuss abolition	US President Ronald Reagan and Soviet President Mikhail Gorbachev meet at Reykjavik, Iceland, where they seriously discuss the possibility of achieving nuclear abolition.
8 Dec. 1987	Intermediate-range missiles banned	The Soviet Union and US sign the Intermediate-Range Nuclear Forces Treaty to eliminate all land-based missiles held by the two states with ranges between 300 and 3,400 miles.
10 July 1991	South Africa joins Non-Proliferation Treaty	South Africa accedes to the Non-Proliferation Treaty. The government claims to have made six nuclear weapons and to have dismantled them all.
15 Dec. 1995	Southeast Asia becomes nuclear-free	The nations of Southeast Asia create a nuclear-weapon-free zone stretching from Burma in the west, the Philippines in the east, Laos and Vietnam in the north, and Indonesia in the south.
11 April 1996	Africa becomes a nuclear-free zone	Officials from 43 African nations sign the Treaty of Pelindaba in Egypt establishing an African nuclear-weapon-free zone and pledging not to build, test, or stockpile nuclear weapons.
1 June 1996	Ukraine becomes a nuclear-free state	Ukraine becomes a nuclear-weapon-free state after transferring the last inherited Soviet nuclear warhead to Russia for destruction. Its president calls on other nations to follow its path.
8 July 1996	World Court says nuclear weapons illegal	The International Court of Justice hands down an advisory opinion in which it found that the threat or use of nuclear weapons would generally be contrary to international law.

24 Sept. 1996	Total nuclear test ban is signed	The Comprehensive Nuclear Test Ban Treaty opens for signatures at the United Nations. China, France, the UK, Russia and the US all sign the treaty. India says it will not sign the treaty.
27 Nov. 1996	Belarus removes its last nuclear missile	Belarus turns its last nuclear missile over to Russia for destruction. It joins Ukraine and Kazakhstan as former Soviet republics that have given up all their nuclear arms.
May 1998	India and Pakistan conduct nuclear tests	India conducts three underground nuclear tests, its first in 24 years. One is a thermonuclear weapon. Later in May, Pakistan tests six nuclear weapons in response to India's tests.
9 October 2006	North Korea conducts nuclear test	The North Korean government announces that it has successfully conducted a nuclear test, becoming the eighth country in the world to do so. It provokes international condemnation.
30 April 2007	ICAN is launched internationally	The International Campaign to Abolish Nuclear Weapons is founded in Australia. It calls for the immediate start of negotiations on a treaty to prohibit and eliminate nuclear weapons.
4–5 March 2013	Norway hosts first humanitarian conference	The Norwegian government hosts the first-ever intergovernmental conference to examine the humanitarian impact of nuclear weapons, bringing together diplomats from 128 states.
14 February 2014	Mexico conference calls for ban	The chair of the Second Conference on the Humanitarian Impact of Nuclear Weapons, held in Mexico, concludes that the time has come for a diplomatic process to ban nuclear weapons.

9 December 2014	Austria issues landmark pledge	As host of the Vienna Conference on the Humanitarian Impact of Nuclear Weapons, Austria issues a landmark pledge to stigmatize, prohibit and eliminate nuclear weapons.
27 March 2017	Nuclear ban treaty negotiations begin	At the United Nations, the overwhelming majority of the world's governments begin negotiations on a treaty to prohibit nuclear weapons, leading towards their total elimination.
7 July 2017	UN adopts nuclear weapon ban treaty	Following weeks of intensive negotiations, two-thirds of the world's nations vote to adopt the landmark UN Treaty on the Prohibition of Nuclear Weapons.

39 Bostrom, N., Dafoe, A., & Flynn, C. (2017). *Policy desiderata in the development of superintelligent AI* (3.7). Retrieved from Future of Humanity Institute, Oxford University website: nickbostrom.com/papers/aipolicy.pdf

40 Bostrom, N. (2016). *Superintelligence: Paths, dangers, strategies.* & Russell, S., Dewey, D., & Tegmark, M. (n.d.). *Research priorities for robust and beneficial artificial intelligence.* Retrieved from futureoflife.org/data/documents/research_priorities.pdf

41 Clark, J. 2016. *Who should control our thinking machines?* Bloomberg. Retrieved from: www.bloomberg.com/features/2016-demis-hassabis-interview-issue

42 Müller, Vincent C. and Bostrom, Nick (forthcoming 2014), 'Future progress in artificial intelligence: A Survey of Expert Opinion, in Vincent C. Müller (ed.), Fundamental Issues of Artificial Intelligence (Synthese Library; Berlin: Springer).

43 Armstrong, S., Bostrom, N., & Shulman, C. (2016). Racing to the precipice: a model of artificial intelligence development. AI & Society, 1–6.

According to the World Population Clock, there are almost 7.5 billion people living on the planet today. This number continues to increase at rapid rates, as death rates are dwarfed by birth rates. Only 200 years ago, the population was one billion, approximately seven times smaller, as stated in Geohive. Likewise, it is predicted that the population will surpass 10 million by 2100.

44 Bostrom, N. (2016). *Superintelligence: Paths, dangers, strategies.*

45 Bostrom, N. (2016). *Superintelligence: Paths, dangers, strategies.*

46 Russell, Sturart, Dewey, Daniel, Tegmark, Max: Retrieved from: *futureoflife.org/data/documents/research_priorities.pdf*

47 Buhr, S. (2017, July 28). CRISPR'd human embryos doesn't mean designer babies are around the corner. Retrieved from *techcrunch.com/2017/07/28/crisprd*-human-embryos-doesnt-mean-designer-babies-are-around-the-corner/

48 Baltimore, D., Berg, P., Botchan, M., Carroll, D., Charo, R. A., Church, G., ... Yamamoto, K. R. (2015). A prudent path forward for genomic engineering and germline gene modification. Science, 348(6230), 36. *doi.org/10.1126/science.aab1028*

49 Regalado, A. (n.d.). Top U.S. Intelligence Official Calls Gene Editing a WMD Threat. Retrieved December 3, 2017, from *www.technologyreview.com/s/600774/top*-us-intelligence-official-calls-gene-editing-a-wmd-threat/

50 Baltimore, D., Berg, P., Botchan, M., Carroll, D., Charo, R. A., Church, G., ... Yamamoto, K. R. (2015). A prudent path forward for genomic engineering and germline gene modification. Science, 348(6230), 36. *doi.org/10.1126/science.aab1028*

51 Mars. (n.d.). Retrieved November 19, 2017, from *www.spacex.com/mars*

52 Urban, T. (2015, January 22). *The Artificial Intelligence revolution: Part 1—Wait But Why.* Retrieved from *waitbutwhy.com/2015/01/artificial-intelligence-revolution–1.html*

53 The Global Information Technology Report 2016. (n.d.). Retrieved December 5, 2017, from *www.weforum.org/reports/the*-global-information-technology-report–2016/

54 Colvile, R. (2017). Getting better government. In *The great acceleration: How the world is getting faster, faster* (p. 231). Bloomsbury.

55 ibid., p. 201

56 The Fourth Industrial Revolution, by Klaus Schwab. (n.d.). Retrieved November 14, 2017, from *www.weforum.org/about/the*-fourth-industrial-revolution-by-klaus-schwab/

57 Schwab, Klaus. The Fourth Industrial Revolution (pp. 68–69). World Economic Forum. Kindle Edition.

58 The Fourth Industrial Revolution: what it means and how to respond. (n.d.). Retrieved December 5, 2017, from *www.weforum.org/agenda/2016/01/the-fourth-industrial-revolution-what-it-means-and-how-to-respond/*

59 Khan, S. (n.d.). *Let's teach for mastery—not test scores.* Retrieved from *www.ted.com/talks/sal_khan_let_s_teach_for_mastery_not_test_scores*

60 Spence, M., Brynjolfsson, E., & McAfee, A. (2014, June 4). New World Order. *Foreign Affairs,* (July/August 2014). Retrieved from *www.foreignaffairs.com/articles/united-states/2014-06-04/new-world-order*

61 Spence, M., Brynjolfsson, E., & McAfee, A. (2014, June 4). New World Order. *Foreign Affairs,* (July/August 2014). Retrieved from *www.foreignaffairs.com/articles/united-states/2014-06-04/new-world-order*

62 Science AMA Series: Stephen Hawking AMA Answers! • r/science. (n.d.). Retrieved December 6, 2017, from *www.reddit.com/r/science/comments/3nyn5i/science_ama_series_stephen_hawking_ama_answers/*

63 Polman, P. (n.d.). Business, society, and the future of capitalism | McKinsey & Company. Retrieved December 6, 2017, from *www.mckinsey.com/business-functions/sustainability-and-resource-productivity/our-insights/business-society-and-the-future-of-capitalism*

64 Friedman, Thomas L.. Thank You for Being Late: An Optimist's Guide to Thriving in the Age of Accelerations (Kindle Location 4625). Farrar, Straus and Giroux. Kindle Edition.

65 Friedman, Thomas L.. Thank You for Being Late: An Optimist's Guide to Thriving in the Age of Accelerations (Kindle Location 53). Farrar, Straus and Giroux. Kindle Edition.

66 Friedman, Thomas L.. Thank You for Being Late: An Optimist's Guide to Thriving in the Age of Accelerations (Kindle Locations 4671–4672). Farrar, Straus and Giroux. Kindle Edition.

67 Friedman, T. L. (2016, February 3). Opinion | Social Media: Destroyer or Creator? *The New York Times.* Retrieved from *www.nytimes.com/2016/02/03/opinion/social-media-destroyer-or-creator.html*

68 Friedman, Thomas L.. Thank You for Being Late: An Optimist's Guide to Thriving in the Age of Accelerations (Kindle Locations 4758–4759). Farrar, Straus and Giroux. Kindle Edition.

69 Tapscott, Don; Tapscott, Alex. Blockchain Revolution: How the Technology Behind Bitcoin Is Changing Money, Business, and the World (Kindle Locations 3617–3618). Penguin Canada. Kindle Edition.

70 Haig, S. (2017, June 18). Bitpesa CEO Claims Bitcoin-Based Remittance Companies Have Reduced Costs by 75%. Retrieved December 16, 2017, from *news.bitcoin.com/bitpesa-ceo-claims-bitcoin-based-remittance-companies-have-reduced-costs-by–75/*

71 State of Connectivity 2016: Using Data to Move Towards a More Inclusive Internet | Facebook Newsroom. (n.d.). Retrieved December 8, 2017, from *newsroom.fb.com/news/2017/02/state-of-connectivity–2016–using-data-to-move-towards-a-more-inclusive-internet/*

72 Literacy rates are rising, but women and girls continue to lag behind | United Nations Educational, Scientific and Cultural Organization. (n.d.). Retrieved December 8, 2017, from *www.unesco.org/new/en/media-services/single-view/news/literacy_rates_are_rising_but_women_and_girls_continue_to_l/*

73 *The Innovator's Dilemma: When New Technologies Cause Great Firms to Fail*, generally referred to as *The Innovator's Dilemma*, first published in 1997, is the most well-known work of the Harvard professor and businessman Clayton Christensen. In the book, Christensen demonstrates how successful, outstanding companies can do everything "right" and yet still lose their market leadership—or even fail—as new, unexpected competitors rise and take over the market.

74 Atkinson, A. (2015). The way forward. In *Inequality* (p. 307). Cambridge, MT: Harvard University Press.

75 How the Fake News Crisis of 1896 Explains Trump—The Atlantic. (n.d.). Retrieved December 8, 2017, from *www.theatlantic.com/technology/archive/2017/01/the-fake-news-crisis–120-years-ago/513710/*

76 Twenge, J. M., & Maby, M. (2017). *iGen: Why today's super-connected kids are growing up less rebellious, more tolerant, less happy—and completely unprepared for adulthood—and what that means for the rest of us*. London: Atria Books.

77 ibid.

78 Escaping the hamster wheel. (n.d.). Retrieved December 16, 2017, from *www.ethz.ch/en/news-and-events/eth-news/news/2017/07/escaping-the-hamster-wheel.html*

79 Csorba, E. (2016, March 22). *4 things Millennials need to navigate the Fourth Industrial Revolution.* Retrieved from *www.weforum.org/agenda/2016/03/4-things-millennials-need-to-navigate-the-fourth-industrial-revolution*

80 Urban, T. (2017, April 20). *Neuralink and the brain's magical future—Wait But Why.* Retrieved from *waitbutwhy.com/2017/04/neuralink.html*

81 Green, L., Brady, D., Ólafsson, K., Hartley, J., & Lumby, C. (2011). *Risks and safety for Australian kids on the Internet.* Retrieved from *www.cci.edu.au/reports/AU-Kids-Online-Survey.pdf*

82 Pornography and the male sexual script: An analysis of consumption and sexual relations. (n.d.). Archives of sexual behaviour, 8–11. doi:10.1007/s10508–014–0391–2.

83 Pizzol, D., Bertoldo, A., & Foresta, C. (2016). Adolescents and web porn: A new era of sexuality. *International Journal of Adolescence Medicine and Health, 28*(2), 169. doi: 10.1515/ijamh–2015–0003

84 Donevan, M., & Mattebo, M. (2017). The relationship between frequent pornography consumption, behaviours, and sexual preoccupancy among male adolescents in Sweden. *Sexual and Reproductive Healthcare, 12*, pp.82–87.

85 Soeters, K., & van Schaik, K. (2006) *"Children's experiences on the internet"*, New Library World, Vol. 107 Issue: 1/2, pp.31–36, doi.org/10.1108/03074800610639012

86 Rideout, V. (2001). *Generation X: How young people use the internet for health information.* Retrieved from Henry J. Kaiser Family Foundation website: *kaiserfamilyfoundation.files.wordpress.com/2001/11/3202–genrx-report.pdf*

87 RSPH. (n.d.). Instagram ranked worst for young people's mental health. Retrieved December 8, 2017, from *www.rsph.org.uk/about-us/news/instagram-ranked-worst-for-young-people-s-mental-health.html*

88 New Vision for Education—Unlocking the Potential of Technology. (n.d.). Retrieved December 13, 2017, from *widgets.weforum.org/nve–2015/*

89 Future of Architects: Extinction or Irrelevance. (2017, June 6). Retrieved December 13, 2017, from *www.di.net/articles/future-architects-extinction-irrelevance/*

90 ibid.

91 Statistics | Malawi | UNICEF. (n.d.). Retrieved December 13, 2017, from *www.unicef.org/infobycountry/malawi_statistics.html*

92 Kharsany, A. B. M., & Karim, Q. A. (2016). HIV Infection and AIDS in Sub-Saharan Africa: Current Status, Challenges and Opportunities. *The Open AIDS Journal, 10*, 34–48. doi.org/10.2174/1874613601610010034

93 Girl Up | Uniting Girls to Change the World | United Nations Foundation. (n.d.). Retrieved December 13, 2017, from *girlup.org/*

94 Population. (2015, December 14). Retrieved December 13, 2017, from *www.un.org/en/sections/issues-depth/population/*

95 Uwihanganye, A. (2017, July 6). *Africa must invest heavily in youth, if we are to lead in innovation*. Retrieved from *www.independent.co.ug/africa-must-invest-heavily-youth-lead-innovation/*

96 7 Architects Designing a Diverse Future in Africa. (2015, February 26). Retrieved December 12, 2017, from www.archdaily.com/603169/7-architects-designing-a-diverse-future-in-africa/

97 Diamandis, Peter H.. Abundance: The Future Is Better Than You Think (p. 40). Free Press. Kindle Edition.

Note: the quote with this reference from Matt Ridley is taken from his article, *Forgwt dollars, cowrie shells or gold", Sunday Times* May 16, 2010.

98 *www.ted.com/talks/dan_pallotta_the_dream_we_haven_t_dared_to_dream*

99 Hwang, V. W., & Horowitt, G. (2012). Conclusion. In *The rainforest: The secret to building the next Silicon Valley*.

100 Harari, Y. N. (2016). The human spark. In *Homo deus: A brief history of tomorrow* (p. 157). New York, NY: Signal.

101 ibid. p. 192.

102 Ibid. p. 198.

103 Csorba, E. (2016, March 22). *4 things Millennials need to navigate the Fourth Industrial Revolution.* Retrieved from www.weforum.org/agenda/2016/03/4-things-millennials-need-to-navigate-the-fourth-industrial-revolution

104 Mangule, I. (2015). *E-democracy in action.* Retrieved from Nordic Council of Ministers website: www.kogu.ee/wp-content/uploads/2015/10/E-democracy-in-Action_case-studies-from-Estonia-Latvia-Finland_2016.pdf

105 ibid.

106 Hani, D., & Kovce, P. (2016). People unwilling to work are sick. In *Voting for Freedom: The 2016 Swiss Referendum on Basic Income: A Milestone in the Advancement of Democracy* (p. 55). Basel: First World Development.

107 Häni, D., & Kovce, P. (2016) *Voting for Freedom: The 2016 Swiss Referendum on Basic Income: A Milestone in the Advancement of Democracy* (Kindle Locations 437–440). First World Development. Kindle Edition.

108 Häni, D., & Kovce, P. (2016) *Voting for Freedom: The 2016 Swiss Referendum on Basic Income: A Milestone in the Advancement of Democracy* (Kindle Locations 2132–2138). First World Development. Kindle Edition.

109 Say, M. (2017, July 10). *How Elon Musk Proved That Thought Leadership Is The New Patriotism.* Retrieved from www.forbes.com/sites/groupthink/2017/07/10/how-elon-musk-proved-that-thought-leadership-is-the-new-patriotism/#26cce6a32440

110 *The Zuckerberg manifesto: How he plans to debug the world.* (2017, February 17). Retrieved from www.dailymail.co.uk/sciencetech/article-4232688/The-Zuckerberg-manifesto-How-plans-debug-world.html#ixzz4qh7fvf3P

111 Spend a moment and watch the lovely movie *The Intern* and empathize with Robert de Niro's lovely character, Ben Whittaker, as he finds his way in the new world and brings meaning to all who meet him at a hip new software company.

112 Schwab, Klaus. *The Fourth Industrial Revolution* (p. 106). World Economic Forum. Kindle Edition.

113 Susskind, R. E., & Susskind, D. (2017). The grand bargain. In *The future of the professions: How technology will transform the work of human experts* (p. 4). Oxford: Oxford University Press.

114 Susskind, R. E., & Susskind, D. (2017). The future of the professions. In *The future of the professions: How technology will transform the work of human experts* (p. 304). Oxford: Oxford University Press.

115 Teaching for the Future: Creating the Teaching Profession that 21st-Century Students Deserve. (n.d.). Retrieved from *www.advanc-ed.org/source/teaching-future-creating-teaching-profession–21st-century-students-deserve–0*

116 Barry, B. (2011). Teaching 2030: What we must do for our students and our public schools—now and in the future. New York, NY: Teachers College.

117 ibid. p.89.

118 ibid. p.101.

119 ibid. p.189.

120 See some of the leading thought leaders here:

Education 4.0 ... the future of learning will be dramatically different, in school and throughout life.—Peter Fisk. (2017, January 24). Retrieved from *www.thegeniusworks.com/2017/01/future-education-young-everyone-taught-together/*

Future Education. (n.d.). Retrieved November 12, 2017, from *www.bbc.com/future/tags/futureeducation*

Classroom of 2020: The future is very different than you think. (2012, October 18). Retrieved from *beta.theglobeandmail.com/news/national/education/canadian-university-report/classroom-of–2020–the-future-is-very-different-than-you-think/article4620458/*

Of all of the subjects in this book, the future of education has been the most complex, diffuse and uncertain for me. At once exhilarating and exasperating, those who live in this world are charged with understanding the future and they are conflicted. They recognize that the future requires most of what is 'education' to be blown up. But they also know that the transition is going to be painful and difficult.

121 The Fourth Industrial Revolution: what it means and how to respond. (n.d.). Retrieved December 5, 2017, from *www.weforum.org/agenda/2016/01/the-fourth-industrial-revolution-what-it-means-and-how-to-respond/*

122 Staying Human in the Machine Age: An Interview With Douglas Rushkoff. (2016, June 21). Retrieved December 12, 2017, from *www.rushkoff.com/staying-human-machine-age-interview-douglas-rushkoff/*

123   While some of this is "urban legend" of Mr. Gates, my younger brother Jay worked at Microsoft for over 15 years and can attest to many meetings where this—or similar conversations like this—were had. The intimation of Gates in a meeting was legendary. A deeply intellectual leader and a hands-on software engineer, Gates did not suffer fools. Many rose because of the leadership style—but they were of a certain type.

124   For a really great read on both sides of the story about Martin Shkreli, read this Vanity Fair piece written in 2015:

McLean, B. (n.d.). Everything You Know About Martin Shkreli Is Wrong—or Is It? Retrieved December 18, 2017, from *www.vanityfair.com/news/2015/12/martin-shkreli-pharmaceuticals-ceo-interview*

125   Gallup Inc. (2013). *The state of the American workplace: Employee engagement insights for U.S. business leaders.* Retrieved from Gallup, Inc website: *www.gallup.com/file/services/176708/State%20of%20the%20American%20Workplace%20Report%202013.pdf*

126   Whoever leads in AI will rule the world': Putin to Russian children on Knowledge Day. (2017, September 1) Retrieved from *www.rt.com/news/401731–ai-rule-world-putin/*

Manufactured by Amazon.ca
Bolton, ON